Understanding Thermodynamics

Understanding Thermodynamics

H.C. Van Ness
Distinguished Research Professor of Chemical Engineering
Rensselaer Polytechnic Institute

Dover Publications, Inc.
New York

Published in Canada by General Publishing Company, Ltd., 30 Lesmill Road, Don Mills, Toronto, Ontario.
Published in the United Kingdom by Constable and Company, Ltd.

This Dover edition, first published in 1983, is an unabridged and corrected republication of the work originally published by McGraw-Hill Book Company, N.Y., in 1969.

Manufactured in the United States of America
Dover Publications, Inc., 180 Varick Street, New York, N.Y. 10014

Library of Congress Cataloging in Publication Data

Van Ness, H. C. (Hendrick C.)
Understanding thermodynamics.

Originally published: New York : McGraw-Hill, 1969.
1. Thermodynamics. I. Title.
QC311.V285 1983 536'.7 82-18250
ISBN 0-486-63277-6

Preface

This short book on the subject of thermodynamics is based on a series of lectures I gave for the possible benefit of some 500 sophomore engineering students at Rensselaer Polytechnic Institute during the spring term of 1968. These lectures were not in any sense meant to replace a textbook, nor were they intended to cover the same ground in the same way as a textbook. I intended them to supplement the textbook, and it was my primary hope that they might help the student over the very difficult ground characteristic of the early stages of an initial course in thermodynamics.

I offer this material in print for the same reason. It falls in the classification of a visual aid, though of a very old-fashioned but perhaps still not outmoded type. It is intended not for experts, but for students. I have left rigor for the textbook and from the very beginning have directed my efforts here toward showing the plausibility and usefulness of the basic concepts of the subject.

H. C. Van Ness

Contents

Understanding Thermodynamics

1

Energy Conservation— The First Law of Thermodynamics

What is thermodynamics? Very briefly, it is the study of energy and its transformations. We can also say immediately that all of thermodynamics is contained implicitly within two apparently simple statements called *the First and Second Laws of Thermodynamics.* If you know anything about these laws, you know that they have to do with energy—the first, explicitly, and the second, implicitly. The First Law says that energy is conserved. That's all; you don't get something for nothing. The Second Law says that even within the framework of conservation, you can't have it just *any* way you might like it. If you think things are going to be perfect, forget it. The Second Law invokes a quantity called *entropy*, something that is not part of our experience, so we'll let it go for a time and consider first the First Law. There is a certain logic in taking up the first things first, and furthermore it allows us to deal with something we all know about, namely, energy.

What is energy? One might expect at this point a nice clear, concise definition. Pick up a chemistry text, a physics text, or a thermodynamics text, and look in the index for "Energy, definition of," and you find no such entry. You think this may be an oversight; so you turn to the appropriate sections of these books, study them, and find them to be no help at all. Every time they have an opportunity to define energy, they fail to do so. Why the big secret? Or is it presumed you already know? Or is it just obvious?

For the moment, I'm going to be evasive too, but I'll return to the question. Whatever it *is*, one thing we know about energy is that it is conserved. That's just another way of saying that we believe in the First Law of Thermodynamics. Why do we believe in it? Certainly no one has proved it. On the other hand, no one has been able to find anything wrong with it. All we know is that it has always worked in every instance where it has been applied, and we are happy with it simply because it works. Why does it work? We haven't the faintest idea; it's just a miracle of nature. The conservation law is a *description* of how nature works, not an *explanation*. Fortunately that's all we need.

Although we do not know why it works, we do know how it works. Any conservation law says that something doesn't change, and any use of the law just involves accounting. We know there is a fixed amount of something, and we need merely find the various pieces that add up, or account for, the total. To give you an idea of how this is done, I am going to tell a ridiculous story. I've stolen the idea of this story from Richard Feynman, Nobel Prize-winning physicist and professor at the California Institute of Technology.

His "Lectures on Physics"[1] should be studied by every serious student of science and technology.

It is the story of 37 sugar cubes, a small boy, and his mother. To set the scene, I will ask you to imagine the boy's room at a corner of a house in rural surroundings. The room has two windows, one facing west and the other facing north. For identification, we will call them window W (for west) and window Q (sorry about that). It happens that window W overlooks a small pond. The boy (perhaps his name is Dennis) plays in this room, and his mother looks in from time to time. One day he asks his mother for some blocks to play with. She has no blocks, but she decides that sugar cubes will do. So she gives him 37 sugar cubes and tells him he is not to eat any or he'll be punished. Each time she returns to the room she counts the sugar cubes lying around, and they total 37; so all is well. But one day she counts and finds only 35. Now Dennis points to an old cigar box he plays with, and his mother starts to open it. But Dennis screams and says, "Don't open the box". The mother, of course, realizes she could open the box anyway, but she's an intelligent, modern mother, and she realizes that this would be a traumatic experience for the boy; so she takes another course.

Later in the day, when she again sees 37 sugar cubes lying about, she weighs the empty box, getting a value of 4.34 oz. She also weighs a sugar cube, getting a value of 0.12 oz. Now the clever lady sets up a formula by which she can check the number of sugar cubes:

$$\text{No. on floor} + \frac{\text{wt of box} - 4.34 \text{ oz}}{0.12 \text{ oz}} = 37$$

This formula works perfectly for quite a time. The left side always totals 37. But one day it does not. Two sugar cubes

[1] Addison-Wesley Publishing Company, Inc., vol. I, 1963; vol. II, 1964.

are missing. As she ponders this problem, she notices that window W is open. She looks out and realizes that the missing sugar cubes could be dissolved in the pond. This taxes her ingenuity, but she was once a nurse and knows how to test the pond for sugar. So she adds a new term to her formula, obtaining

$$\text{No. on floor} + \frac{\text{wt of box} - 4.3\text{ oz}}{0.12\text{ oz}} + k\,(\text{sucrosity of pond}) = 37$$

and determines the proportionality constant k by tossing a cube into the pond herself.

This fixes up her formula, and again it works perfectly, accounting always for 37 sugar cubes. As she uses the formula, she begins to realize she could make her work easier if she dealt with *changes* in the various terms from one checking of the formula to the next. From this point of view the formula can be written as

$$\Delta(\text{No. on floor}) + \frac{\Delta(\text{wt of box})}{0.12\text{ oz}} + k\,\Delta(\text{sucrosity of pond}) = 0$$

where the symbol "Δ" means *change of*. This equation simply says that if sugar cubes are conserved, the sum of all changes in the number of sugar cubes in various places must be zero. This equation too works perfectly for a long period, but one day it fails. The sum comes out not zero, but −4. Four sugar cubes are missing! This time it doesn't take mother long to notice that *both* windows are open and that she has no term in her equation to account for sugar cubes thrown out through window Q. She does not see any sugar cubes on the ground outside, but she does see several squirrels running about. How can she possibly keep track of all that goes on outdoors? The pond was bad enough, but what about squirrels and who knows what else? Her husband, an electrical engineer, solves her problem by building a detection system at each window that counts the sugar cubes as they fly past, so it is no longer necessary to keep track of what happens outdoors. It is only necessary to record what passes through the walls of the room. The mother revises her formula again to reflect the new accounting procedure:

$$\Delta(\text{No. on floor}) + \frac{\Delta(\text{wt of box})}{0.12 \text{ oz}} + \text{No. passing } W + \text{No. passing } Q = 0$$

Note that we did not put Δ's with the two new terms. They need not be thought of as a *change* in anything. They just represent a number of objects passing a boundary during the interval between checks. In fact, we may as well simplify these terms to read W and Q. We can also transpose them to the other side of the equation; the result is

$$\Delta(\text{No. on floor}) + \frac{\Delta(\text{wt of box})}{0.12 \text{ oz}} = -Q - W$$

You see that we are getting more and more technical, and when this happens, technical terms also begin to appear. We may as well introduce several such terms here. Notice that we have narrowed our attention down to the room and to its walls, i.e., to a small region of space. In technical language, we call the room our *system*, and the walls become its *boundary*. Everything outside the boundary is called the *surroundings*. We would very much like to get rid of the surroundings because of their infinite complexity, but we can't really ignore them. On the other hand, we can make our formula *look* like it deals only with the system. The last form in which we wrote our formula puts the terms that have to do with changes in the system on the left. On the right we have terms to show what passes out of the system, but they are really there to account for changes in the surroundings. By associating them with the boundary of the system we make the appearance of dealing solely with the system. We treat Q and W as quantities, not as changes in anything, but in fact they are there to account for changes in the surroundings. Any conservation law must somehow include both the system and its surroundings. By insisting that we can account for the surroundings by counting at the system boundary, we are in fact adding something new to the content of a conservation law. It is a bit subtle but becomes obvious once pointed out. We do not expect one of Dennis' sugar cubes suddenly to disappear from his room on one side of the world and simultaneously to reappear someplace on the other side of the world, even though the other side is part of the surroundings. Why not? No simple statement of a conservation law excludes this. But it just isn't reasonable; it doesn't make sense. We'll leave it at that. The point is that the system-and-its-boundary formula excludes this possibility. We could also exclude it by insisting that conservation exists between a system and its local surroundings, but then we would have to define "local" as any part of the universe with which the system interacts. Then we would find it necessary to define "interacts," and so on. The beauty of the mathematical system-and-its-boundary formula is that it avoids this chain of verbiage, and this is one of the major advantages of the use of mathematics in the formulation of the laws of science, not that conservation of sugar cubes is a law of science—not yet, at any rate. So let's return to Dennis, his mother, and the 37 sugar cubes.

All is going well, except that Dennis' final sugar cube just entered the surroundings. It's time for a new game, and Dennis' mother dumps a handful of sugar cubes in his cigar box. This time she doesn't even count them. Can she still play the game? She certainly can, and she can even delay the start. All it takes is an initial observation and the setting of the window counters to zero. The formula always works, and when the counter on window Q breaks down, the mother realizes she can use her formula to find out how many sugar cubes are being fed to the squirrels.

Now consider another situation. A friend comes to visit and hands Dennis a bag of jelly beans. The mother doesn't happen to see this, and the friend says nothing about it. Furthermore, Dennis treats the jelly beans as if they were illegal; he never leaves any on the floor, and he won't say what he has. His mother is very curious, but all she knows is that Dennis has *something* in his cigar box. Nevertheless, she decides to try her formula; and it's going to work, because jelly beans come in lumps, and that's essential to her accounting scheme. Remember, she does not know what Dennis has. It is only necessary that she believe in lumps; she doesn't have to *see* them. She *does* have one problem; she does not know the weight of a lump. So her formula must be written

$$\frac{\Delta(\text{wt of box})}{a} = -Q - W$$

How does she get a, the weight of a lump? There's only one way; she must use her formula. So she weighs the box and sets the counters. Then after an interval she reweighs the box, records Q and W; now a is the only unknown in her formula, and she determines its value. After that she can use the formula to check her "law of conservation of lumps." Alternatively, she can use it to calculate any one of the three factors in it from the other two, provided only that she accepts the law of conservation of lumps to be valid.

Perhaps this is all absurdly obvious. If so, we can make it more cryptic by noting that the left-hand member of the mother's formula can be viewed a bit differently. This formula represents no more than a counting scheme, and Q and W represent counts directly. But the left-hand member is a count only indirectly. Clearly, the number of counts that it represents is given by the change in some *function of* the weight of the box. Thus the equation may equally well be written

$$\Delta[f(\text{wt of box})] = -Q - W$$

where $f(\text{wt of box}) = \dfrac{\text{wt of box}}{a}$

Here we know precisely the nature of the function f(wt of box), and by experiment we have established the value of a, the only adjustable parameter in it. However, we can imagine more complex situations where the function depends on properties of the box other than weight (perhaps on its electric charge or its permeability to x-rays). Moreover, the nature of the function may be far from simple. Thus we begin to see how a conservation law can become both difficult and abstract.

The law of conservation of energy is inherently more difficult and abstract because it does not deal with the conservation of lumps. Energy does not come in uniform lumps. This law proclaims the conservation of a number which does not represent any particular *thing*. Let's examine this in detail. How is energy conservation similar to and different from conservation of sugar cubes and jelly beans? They are alike in that the formulas which describe their conservation are mathematically similar; that is, the formulas include terms that account for changes in both the system and its surroundings. Moreover, their simplest and most convenient expression is given in terms of changes which occur within the system and in terms of quantities which pass the boundary of the system. This also requires that conservation be *local*. Thus we can write the same equation for energy conservation as we did for the conservation of lumps. It is analogous to the case in which we never saw the lumps, for nobody has ever claimed to *see* energy. The energy of a system is no more evident than jelly beans enclosed in a cigar box. So our conservation formula has the form

$$\Delta(\text{energy of system}) = -\text{energy out by } Q - \text{energy out by } W$$

The energy of the system is presumed to be some function of the measurable properties of the system, just as the number of jelly beans was a function of the weight of the box. But we can't weigh energy, and the functional relationship is not known ahead of time. We can only *guess* of what property the energy of the system is a function. So we guess that it may be a function of temperature, pressure, composition, magnetization, etc. We really don't know so we'll leave it indefinite by writing

$$\text{Energy of system} = U(T,P,\text{etc.})$$

where we call U the *internal energy function*, and the parentheses show of what property it is a function. Our conservation formula is now written

$$\Delta[U(T,P,\text{etc.})] = -Q - W$$

The notation is often simplified still further so that we usually have

$$\Delta U = -Q - W$$

and we get careless with our terminology and call U simply the *internal energy*, as though we know exactly what we're talking about. But in fact we don't, and U is known only as a function of other things.

Q and W are terms representing energy passing the system boundary, not just by different windows, but by different modes. They are called *heat* and *work*, respectively, and both words have a special technical meaning. Both can give us all sorts of trouble, but for the moment let's assume we know all about them and can measure them.

You may be thinking that I have somehow *derived* the equation of energy conservation. Nothing could be further from the truth. I have just written it down. That's all anybody can do. No matter how much is written about this equation in thermodynamics texts, no matter how many fancy diagrams are drawn, no matter how confused the issue is made by mathematical manipulations, if you look carefully, you will find in the end that the author has merely written it down. No fundamental law of science is derivable by any means that we know today. If we could derive such laws, they would not be called fundamental. Then have I *explained* the law of conservation of energy? Again, I have not. I have tried to make the fact that it works seem plausible, but primarily I am trying to show you *how* it works, and there is a little way to go yet.

One thing about my equation may be bothering you. It is written with minus signs on both Q and W. The origin of these minus signs lies in the fact that Dennis could throw sugar cubes only *out of* the system. Had sugar cubes somehow come only *into* the system, both signs would be plus. However, the equation is usually written

$$\Delta U = +Q - W$$

This is just an accident of history. The first applications of thermodynamics were made to heat engines, devices which take *in* heat and put *out* work. The signs merely reflect a decision on the part of the founding fathers to make heat *in* and work *out* positive quantities for their favorite device. You can write it any way you want, that is, $+Q + W$, $-Q - W$, $-Q + W$, or $+Q - W$. All that is required is consistency in ascribing signs to your *numerical* values of Q and W. We will henceforth follow the crowd and write

$$\Delta[U(T,P,\text{etc.})] = Q - W$$

How is this equation to be used? For engineering purposes we want to use it to calculate either Q or W, or even both Q

and W if we can find a second equation connecting Q and W. But how can we use it without knowing the functional relation $U(T,P,\text{etc.})$? How are we to get numbers for this function? How are we even going to find out what U is a function of?

The answer to the last question is easiest. It involves the notion of *state.* We say that the internal state of a system is fixed when none of its measurable properties changes any more. Then the problem is to find what measurable properties we need to establish at arbitrary values in order to fix the state of a system. This is one of the major complications of thermodynamics—to know what the variables are. The only way to find out is by experiment. The internal energy is presumed to be a function of the same variables as is the volume.

Having established the variables, say temperature T and pressure P, how do we get the relationship between U and these variables? This is the second major complication of thermodynamics. We find in any ultimate analysis that we must use our equation of energy conservation. This may seem incredible; after all, the use of the equation

$$\Delta[U(T,P)] = Q - W$$

is to find Q or W. How can we use it to calculate values for $U(T,P)$ and Q or W at the same time? The secret is that we don't do both *at the same time.* We play the game forward and backward, but at different times, just as Dennis' mother did when she used her equation backward to determine the weight of a jelly bean or lump that she never saw. Having done that, she could subsequently use her equation forward to check on the conservation of lumps or to find the number of lumps in the pond or the number fed to the squirrels. The same thing holds true for the energy equation, except that the process is more complicated because not only do we never see the energy, it does not come in lumps.

In the laboratory we set up a small system and make

changes in it, measuring T and P and Q and W. From these it is possible to deduce $U(T,P)$ for various values of T and P within an additive constant. In use we always have $\Delta[U(T,P)]$ so that the constant drops out. We can put down $U(T,P)$ in the form of a graph, a table, or an equation. But we must *have* such information, and it must ultimately come from experiment. Moreover, we must have $U(T,P)$ or $U(T,P,\text{etc.})$ for the particular kind of system we wish to deal with. Given this information, we may apply the energy formula

$$\Delta[U(T,P,\text{etc.})] = Q - W$$

to any process involving the same kind of system, and it is in no way limited to just those processes used to determine $U(T,P,\text{etc.})$. Any such limitation would make it of no use at all.

Let us say that we know $U(T,P,\text{etc.})$ and now apply our conservation formula to many different processes. We find time after time that it checks out, that it works. Then one day it doesn't. What to do? We do just what Dennis' mother did. We look for sugar cubes under the rug, in the pond, or in any place we had not considered before. We notice that our system changed its elevation. Maybe that changes its energy. Sure enough, a bit of experimentation shows that we can devise a potential energy function which fixes our formula—for a time. We go through the whole business again and find we need a kinetic energy function when the system has velocity. So we add terms to our formula as follows:

$$\Delta[U(T,P,\text{etc.})] + \Delta[\text{PE}(z)] + \Delta[\text{KE}(u)] = Q - W$$

Fortunately, the two new functions are known explicitly in terms of measurable properties; thus

$$\text{Potential energy function} = \text{PE}(z) = mgz$$

$$\text{Kinetic energy function} = \text{KE}(u) = \tfrac{1}{2}mu^2$$

where z = elevation
m = mass
g = acceleration of gravity
u = velocity

Thus

$$\Delta[U(T,P,\text{etc.})] + mg\,\Delta z + \tfrac{1}{2}m\,\Delta u^2 = Q - W$$

And so it goes. Whenever our equation does not work, we can fix it up with a new term. Others may object that this isn't fair and accuse us of deciding arbitrarily that the law of conservation of energy *is* valid and of being *determined* to make it work. They claim that we doctor it up so that everything comes out all right. They would, of course, be correct if we ever added a term called "unaccounted for" or "lost." That would spoil it all. But it turns out that every time we add a new term to our equation, we're also able to say how to evaluate it from measurable parameters. This sort of doctoring is completely justified. Can we do the same thing with a law of conservation of sugar cubes? The answer is no. What if Dennis stomps on a sugar cube? We still have sugar but no cube. Or he may eat one, and then we don't even have sugar.

Perhaps the ultimate test of our accounting scheme came with the advent of nuclear fission. Energy appears in this case to come from nowhere, but in fact a term provided by Einstein readily maintains the validity of the conservation equation. The new term is a nuclear energy function, and its change is $-c^2\,\Delta m$, where c is the velocity of light and Δm is the change in mass of the system. The minus sign is necessary because Δm is negative; the mass of the system decreases. Our equation then becomes

$$\Delta[U(T,P,\text{etc.})] + mg\,\Delta z + \tfrac{1}{2}m\,\Delta u^2 - c^2\,\Delta m = Q - W$$

When we're all done, what do we have? We have an equation which is said to give *mathematical* expression to the law of conservation of energy. But how else could this law be expressed except mathematically? Every form of energy we have discussed is known *only* as a function of other variables, and I have been careful to say internal energy *function*, potential energy *function*, etc. Functions are *pencil-and-paper* constructs. I can't show you a function that has any other substance, and that is why I can't show you a chunk of energy or why I can't define it or tell you what it is. It is just mathematical or abstract or just a group of numbers. Thus we have no energy meters, no device we can stick into a system which will record its energy. The whole thing is man-made.

What we have is a scheme with a set of rules. The scheme involves only changes in the energy functions. It is set up this way because we have no way to calculate absolute values of our energy functions. The remarkable thing about this scheme is its enormous generality. It applies equally to the very small and to the very large; it applies over any time interval, short or long; it applies to living matter as well as to dead. It applies in the quantum-mechanical and relativistic realm as well as in the classical. It just plain works. We can never be absolutely sure that it will always work, but we are sufficiently confident so that with it we make all sorts of predictions, and that is its use.

2

The Concept of Reversibility

In Chap. 1 I talked about energy and its conservation from a very general point of view. I tried to show how we go about the business of accounting for energy. We ended with a scheme and a set of rules—a *formalism*. Now this formalism is something we have created to serve our own ends, but it has built-in limitations which derive from the fact that energy is not a thing, like a sugar cube or a jelly bean. My point is that this particular formalism may not be appropriate for just any old system we might select in applications of the law of conservation of energy. The law itself is, of course, always right, but the particular formalism used to express it may present difficulties if the system is not selected with care.

For example, consider a block of steel acted upon by a force F so that it slides at uniform velocity over another piece of steel, as shown in Fig. 2-1. Let us apply our formal-

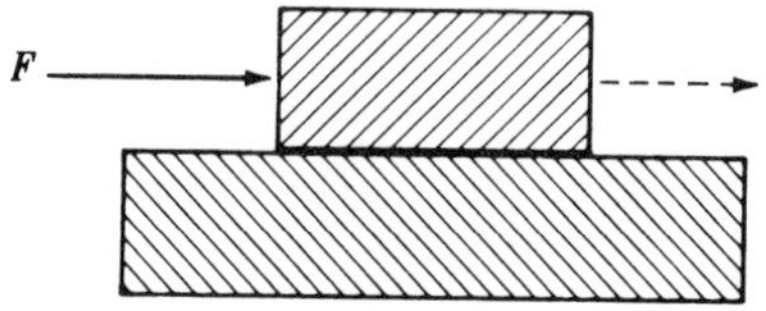

Figure 2-1

ism to the first block taken as the system. This block is shown in Fig. 2-2 along with the forces that act on it along

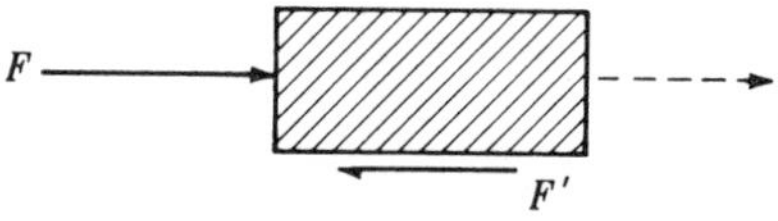

Figure 2-2

the line of its motion. Since the velocity of the block is uniform, the frictional force F' must be equal and opposite to the applied force F. Thus

$$F - F' = 0$$

and there is no net force on the block. Since the work done on the block is equal to the net force times the distance through which it moves, we conclude that the net work done on the block is zero. The First Law of Thermodynamics as we have expressed it may be written for the block as

$$\Delta U = Q - W$$

where ΔU is the internal energy change of the block, Q is the heat added to the block, and W is the net work done by the block. But $W = 0$, and therefore

$$\Delta U = Q$$

This result says that the internal energy change of the block is equal to the heat transferred to the block from the surroundings. Remember that Q is a term which is included to account for energy changes in the surroundings. However, we call it *heat* because it is energy transferred across the boundary of the system as a result of a temperature difference. Now we know from experience that the temperature of the sliding block increases, and this increases its internal energy. According to our equation this implies a transfer of heat to the block from the surroundings, which would then necessarily be at a higher temperature than the block. But there is no mechanism by which the temperature of the surroundings is raised above that of the system (the moving block). Thus if Q has the significance we have attached to it, the equation must be wrong, and if the equation is correct, we must redefine Q. The formalism we have developed simply does not apply to the *particular* system chosen in this example. Our choice was such as to make the system boundary the site of a transformation of energy. The mechanism is friction, and the friction occurs at the boundary separating the system from the surroundings. This always leads to embarrassment.

We can either abandon the formalism or select a new system. Usually, we take the latter course and pick a different system. In this case we may take both blocks as the system. This serves to put the friction *inside* the system. Figure 2-3 shows our new system and the force F acting on it. Both block 1 and block 2 experience changes in internal

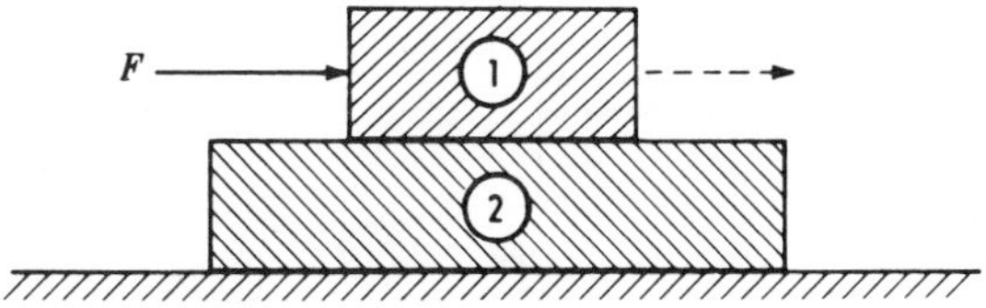

Figure 2-3

energy, and we write our First Law equation as

$$\Delta U_1 + \Delta U_2 = Q - W$$

The work W is simply the work done by the force F as it moves a distance Δs. This work is negative (done on the system); thus we write $W = -F\,\Delta s$. Therefore,

$$\Delta U_1 + \Delta U_2 = Q + F\,\Delta s$$

In this equation Q is not heat transfer between the blocks, but heat transfer from both blocks to their surroundings. We now have an entirely proper equation and have avoided any embarrassment. On the other hand, we cannot by thermodynamics alone evaluate ΔU_1, ΔU_2, or Q. Thermodynamics just tells us what terms we need to take into account, and it gives us an equation relating the terms.

With this illustration I've tried to make two major points. First, we must use judgment in the selection of a system if we expect to use the formalism by which the First Law is normally expressed. Second, many problems cannot be solved by thermodynamics alone.

Another apparently simple device encountered endlessly in thermodynamics is the piston-and-cylinder combination, as shown in Fig. 2-4 (we usually consider a gas to be trapped in the cylinder). If embarrassments of the kind encountered with the sliding block are to be avoided here, we must assume that the piston moves in the cylinder without fric-

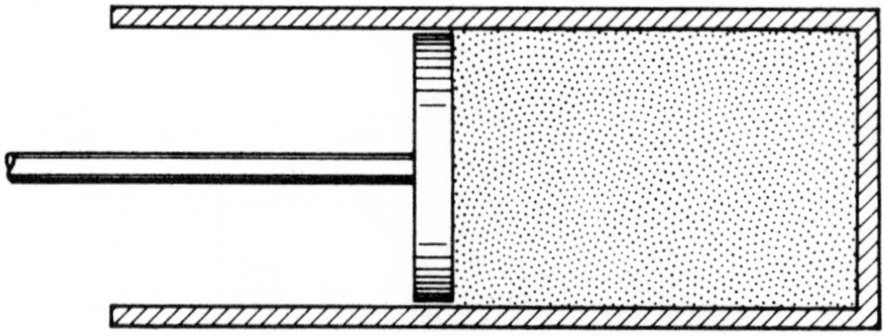

Figure 2-4

tion. This is good for all sorts of trivial applications of thermodynamics, but it can also be used to illustrate a number of very important concepts; we will use it extensively for this purpose.

Imagine that we have such a piston-and-cylinder combination and that a weight w is placed on the piston to hold a gas under compression in the cylinder. This initial state of the system is shown on the left in Fig. 2-5. We will

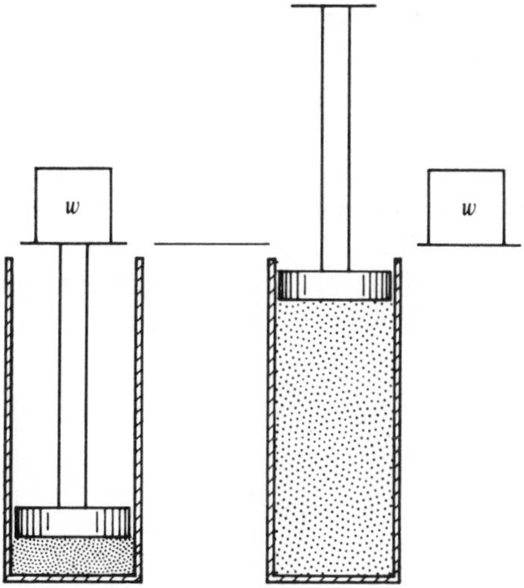

Figure 2-5

assume that the piston is so perfectly lubricated that it can move without friction in the cylinder. In addition, we will assume that the piston and cylinder are constructed of a special material that is a perfect heat insulator. Thus there can be no heat transfer between the gas and its surroundings. Any such process is said to be *adiabatic.*

We wish to use the compressed gas in the cylinder to accomplish useful work, and the question is how to carry out a process so that we may obtain the maximum possible useful work. Raising the weight w will be considered the object of the process and will thus constitute the performance of useful work. Since the initial state as shown in Fig. 2-5 is one of equilibrium, it is clear that the piston will not move unless the weight w is removed from the piston. Imagine that the weight is struck from the side so as to cause it to slide suddenly to an adjacent shelf. The piston, of course, shoots upward and after a period of up-and-down oscillation settles into a final equilibrium position, as shown on the right in Fig. 2-5. On the other hand, the weight w has not been raised, and no purpose has been served by the process. We must do things differently.

We decide to divide the weight into two parts so that we need not remove all of it from the piston at once. Again we start with the piston held in position by the weight w, as shown in Fig. 2-6. This time we slide only $\frac{1}{2}w$ off the piston to an adjacent shelf. The piston again shoots upward and eventually comes to rest in a new equilibrium position,

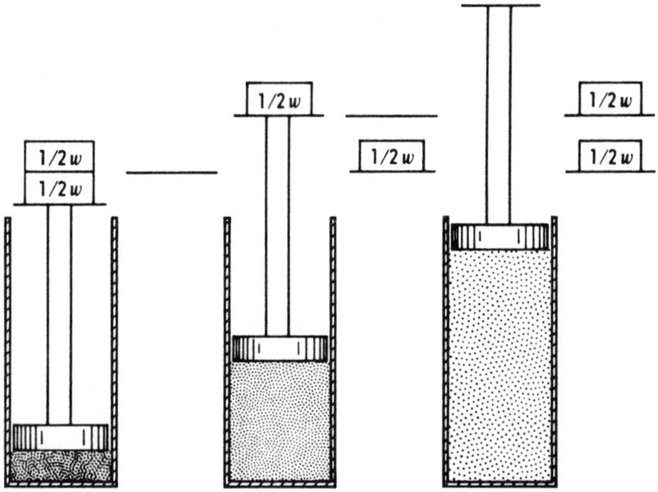

Figure 2-6

but this time it has carried half the weight w with it. This $\frac{1}{2}w$ may now be pushed off to another shelf, and the piston is again free to find a new final position. All three stages of this process are shown in Fig. 2-6. Clearly, a weight has been raised; the process has been of some use. Half of the weight w has been raised somewhere near half the distance of the piston stroke. But is this the best we can do? What if we divide up the weight into much smaller bits and use, for example, a pile of sand? We might imagine flicking grains of sand off the pile one by one and having them stick to a sheet of flypaper. Various stages of this process are depicted in Fig. 2-7. The removal of each grain of sand

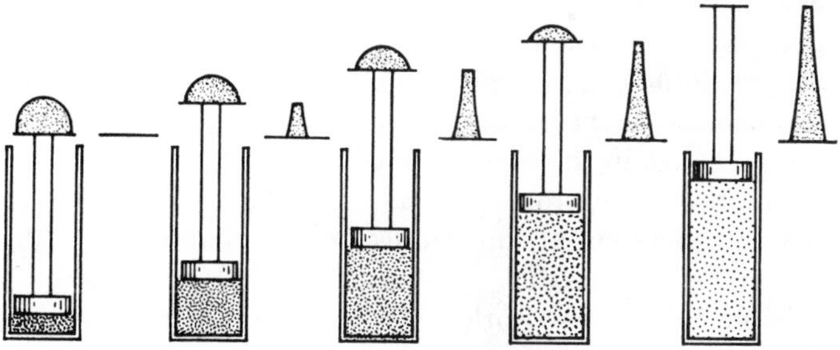

Figure 2-7

from the piston causes very little change in the system. The piston moves but a very small distance at a time, and there is but the slightest oscillation of the piston as it finds a new equilibrium position only a hair above its previous position. Clearly the final grain of sand is carried with the piston almost to the end of its stroke, and in the end we have raised all the weight w (the sand) an average distance of something like half the stroke of the piston.

All we can do to improve on this last process is to use a finer sand and in the limit to make the grains infinitesimal or vanishingly small in size. Then the process is carried out in differential steps, and we can only imagine it. Such an imaginary process would improve but slightly upon the one just described. But it does represent the limit of what can be done by way of improving the process and thus providing the maximum possible work, and as such it is worthy of our close attention.

This imaginary process is called *reversible* because at any point it could be turned around and made to go the other way simply by replacing the infinitesimal grains of sand on the piston. Only one *additional* infinitesimal grain of sand would be needed to start the reverse process. Then the particles previously removed from the piston could be placed back on the piston at exactly the same level they had on the flypaper. In other words, only a differential change in conditions would be required to reverse the process, and then the reverse process would carry the system back to its initial state, leaving only a minute or differential change in the surroundings.

The reversible process is unique, and as such occupies a position of essential importance in thermodynamics. The reason for this is that it represents the *limit* of what is possible in the real world. We cannot even imagine anything better. Moreover, it lends itself to exact mathematical analysis, and this is not true of any other process. Our choice in thermodynamics often is to do calculations for reversible processes or to do no calculations at all. The reason for this is that reversible processes are those for which the forces causing change are almost exactly in balance. Thus the states through which the system passes during a reversible process are for all practical purposes equilibrium states, or more precisely are never removed more than differentially from equilibrium states. The importance of this observation is demonstrated by the following considerations. The work done in raising a weight is given by $\int F\,ds$, where F is the force of gravity on the weight and s is the elevation of the weight above some arbitrary but fixed datum level. Now if we wish to calculate the work done *by the gas* in any of the processes described earlier, the force F must be the gravitational force acting downward on all of the mass supported by the gas at pressure P. This mass includes that of the piston, the piston rod, the pan, the weight w, and the atmosphere above the piston. In the case of the reversible process this force F is never more than minutely out of balance with the force exerted upward on the piston face by the gas and given by the product of pressure and the piston area. Thus, for all practical purposes, $F = PA$ for the reversible process. In addition, the volume change of the gas (the system) is always given by $dV = A\,ds$. Thus $ds = dV/A$, and the work done by the gas is

$$W = \int F\,ds = \int PA\,\frac{dV}{A} = \int P\,dV$$

Thus if we can substitute PA for F, we can calculate the work from knowledge of the system without knowing anything about what happens in the surroundings. This substitution is possible only for reversible processes where the forces are never more than differentially out of balance.

For irreversible processes this substitution is not possible. When a finite weight is removed from the piston in the processes described, the force of gravity acting downward is overbalanced by the gas pressure acting upward by a finite amount, and F does not equal PA again until a new equilibrium position of the piston is reached. Thus PA cannot be substituted in the integral $\int F\,ds$, and it is not possible to calculate the work from a knowledge of the properties of the system. Thus we have the important result that the work done by the system (the gas) is given by $\int P\,dV$ only when the process is reversible; that is,

$$W_{\text{rev}} = \int_{V_1}^{V_2} P\,dV$$

Moreover, this work for an expansion process is the maximum work which the system can produce. If we were to consider the compression of a gas by a piston in a cylinder, we would obtain the same results except that the reversible work would be the minimum work required for compression of the gas. The difference is, of course, that a compression process is carried out by work done *on* the system, whereas an expansion process results in work done *by* the system. In either case the reversible work is a limiting value, i.e., the maximum obtainable when work is produced and the minimum required when work is expended.

If we were to sit down to flick grains of sand one by one off a piston, we would have to be very patient indeed in order to wait out any appreciable change in our system. If the grains of sand were made infinitesimal, any finite process would require an infinite time. This is characteristic of all reversible processes; we must imagine them to proceed infinitely slowly. This is consistent with the fact that they are imagined to be driven by an infinitesimal imbalance of forces.

We assumed our piston-and-cylinder processes to occur without friction for a very good reason. Without this assumption we could not even imagine a reversible process. If there were friction between the piston and the cylinder, the piston would stick, and we would always have to remove a finite amount of sand from the piston before it would move at all. Then it would move in jumps, and the condition of virtual balance of forces which allows us to substitute PA for F would be violated.

The other assumption we made was that the process was adiabatic, i.e., that there was no heat transfer. This assumption was made merely for convenience, so that we could concern ourselves with just the mechanical aspects of the process. It is entirely possible to imagine reversible heat transfer. The driving force for heat transfer is a temperature difference, and for reversible heat transfer we need only imagine that this temperature difference becomes infinitesimal. Thus heat is transferred reversibly when it flows from an object (or a system) at temperature T to another object (or the surroundings) at temperature $T - dT$.

The concept of reversibility is essential to the subject of thermodynamics. Its abstract nature in no way destroys its practical utility, as I hope to demonstrate in the next chapter.

3

Heat Engines

In this book I don't feel constrained to present material in the same order that is found in any standard textbook. My purpose, in fact, is to do it differently. The problem in teaching thermodynamics is that the most difficult, the most confusing, and the least interesting material is presented first, and the early applications are usually at best trivial and at worst misleading. But this material and these examples are just a prelude, and in this chapter and the next I want to get beyond all this so that you can get some idea of the usefulness of thermodynamics.

In my last lecture I discussed the problem of getting work from a compressed gas by means of an adiabatic expansion in a piston-and-cylinder assembly. We found that the maximum work could be obtained only for a reversible expansion. Furthermore, we found that we could calculate the work by $\int P\, dV$ only for such a process. For irreversible processes things are not so easy. Unfortunately, it is a matter of everyday experience that all processes in the real world are irreversible. The watch you are wearing "runs down." An oscillator must have a continuous supply of power or it stops. The earth slows its rotation because of the tides. All living things grow older. You can readily scramble an egg, but the only convenient way to unscramble it is to feed it to a hen, and this presumably requires the expenditure of considerable effort on the part of the hen.

If all real processes are irreversible, why do we spend so much time discussing reversible processes? In the first place, the reversible process represents a *limiting behavior*, i.e., the best that we can hope for. Thus we employ it as a standard against which the performance of real processes may be measured. Secondly, the reversible process is one for which we can readily do the calculations. The alternative is likely to be that we do no calculations at all.

We do not *always* need the assumption of reversibility. There are, in fact, two situations. First, there are problems for which an energy balance can be written that contains only one unknown term. In this case we need not concern ourselves with the question of reversibility; we just solve the equation. Energy is *always* conserved for reversible and irreversible processes alike. Second, there are problems for which the energy balance contains more than one unknown term. Here the assumption of reversibility is usually essential if we wish to do any calculations at all. So we assume reversibility when we are forced to and in that way change the problem into one we can work. But then are we not just working imaginary problems? We are, indeed. But what use is that? There are, in fact, two uses, and thermodynamics splits here into two complementary parts. The first part might be called *theoretical*, because it concerns itself with the properties of materials and their experimental measurement. The second part might be called *applied*, because it concerns itself with the solution of practical problems through use of the properties provided by the first part.

I pointed out earlier that the equation

$$\Delta U = Q - W$$

is used forward (to solve problems) and backward (to provide values of U). I'm talking about the same thing again. Consider the theoretical, or backward, use of this equation. For this we usually use the differential form of the energy equation, $dU = dQ - dW$. Now we are just interested in the property U of some material, say a gas, and in how it is related to other properties such as P, V, and T. We can imagine that the gas (a real gas) undergoes any kind of process we like. Why not imagine the process to be reversible? Then we can substitute $P\,dV$ for dW, and later we will see that in the same way we can substitute $T\,dS$ for dQ; thus we get

$$dU = T\,dS - P\,dV$$

Now we *derived* this equation for a reversible process, but once derived we see that it contains just properties of the system, and so it must not depend on the kind of process considered. What we have really done is to derive an equation for a special case and then conclude that it must be general! Thermodynamics does things backward! We are always generalizing. But once having the equation we can forget our reversible process (which was but a means to an end) and enter the abstract world of pure mathematics. You'd never believe what's done with this poor little equation (and all of it exactly right, because it depends only on mathematics). That's the theoretical half of thermodynamics, and although we will pay some attention to it, we are really more interested in the other half—applied thermodynamics.

What then do we do about irreversibilities in real processes where the problem can't be solved without paying attention to the irreversibilities? You will be horrified to learn that we ignore them! At first, we make believe they aren't there and assume the process is reversible. This allows us to work the problem (incorrectly to be sure), but after we're done we ask how far wrong we are and make corrections to get realistic but approximate answers.

To see how this works, we'll take a specific example—the Otto engine cycle. This cycle is an idealization of what goes on in virtually all gasoline engines. The engine which follows this cycle was constructed by Nikolaus Otto in 1876. Otto did not invent the internal-combustion engine, but he made it practical by devising a cycle by which it could operate in an efficient manner. (Incidentally, it was first used to power an automobile by Karl Benz in 1885, and it was another 10 years before Henry Ford made his first car. Ford's contribution was mass production, which started with the Model T, introduced in 1908. But let's get back to the Otto engine cycle.)

One imagines a piston and cylinder containing air. The piston rod is connected to a crank which drives a shaft and flywheel in rotary motion, but it's the process going on inside the cylinder that interests us, for that is where the action is. We imagine that the piston slides in the cylinder without friction and that all processes are carried out reversibly. The changes which occur in the air within the cylinder are depicted in Fig. 3-1, which shows a graph of air pressure vs. cylinder volume. The piston, of course, goes back and forth within the cylinder, and when the engine is in continuous operation, the air goes through a series of steps which constitute a cycle, as shown in the figure. At point 1 the piston is at its point of maximum withdrawal from the cylinder, which is full of air at about atmospheric pressure. In an actual engine an amount of gasoline vapor is mixed with the air, but in the idealized process used to analyze the cycle this is neglected. The piston is carried into the cylinder during step 1–2 by the inertia of the flywheel, which provides the work W_{in}, and the air is compressed in what is imagined to be a reversible, adiabatic process. When the piston reaches the limit of its stroke at point 2, heat is added to change the conditions of the air to

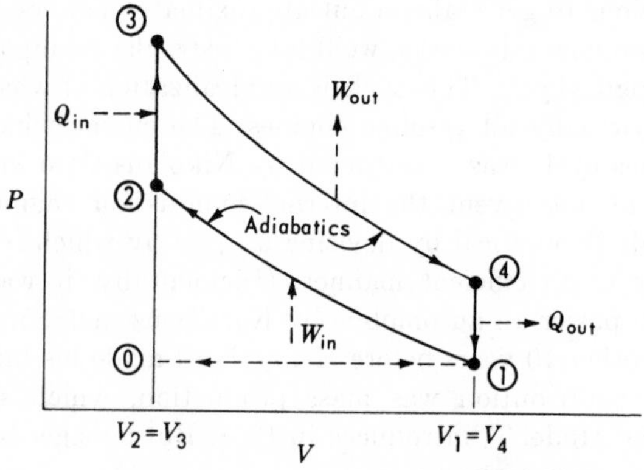

Figure 3-1 Otto engine cycle.

those represented by point 3. This heat addition, called Q_{in}, causes the same change as the combustion of the gasoline in an actual engine. Since this is a rapid process, the heat is imagined to be added very quickly during the instant that the piston remains in its limiting position. Both the pressure and the temperature at point 3 are much higher than at point 2. The gas now expands against the piston during step 3–4, doing work W_{out} in what is imagined to be another reversible, adiabatic process. At point 4 we imagine that the heat Q_{out} is very quickly removed from the gas to restore the conditions of point 1. In an actual engine, the exhaust valve opens at point 4, and the hot gases rush out of the cylinder. In addition, the actual engine goes through an extra cycle (represented by the horizontal dashed line) which produces no work. The purpose of this is to push exhaust gases from the cylinder (step 1–0) and to draw in fresh air and fuel (step 0–1). The net result is to return the system to state 1, just as is done by the single step 4–1 in our idealized cycle.

We now are ready to do a few calculations, but we must first specify certain conditions for our cycle. It is reasonable to assume that air behaves essentially as an ideal gas under the conditions which exist for the cycle, and we will take its heat capacity at constant volume, C_V, as constant at 5 Btu/lb mole–°F. This makes the ratio of heat capacities $\gamma = C_P/C_V = 7/5 = 1.4$. The conditions at point 1 are taken to be $P_1 = 1$ atm and $T_1 = 140$°F (600°R). The compression ratio of the engine is such as to make $P_2 = 15$ atm, and sufficient heat is added in step 2–3 to make $T_3 = 3940$°F (4400°R). Nothing further need be specified to allow calculation of everything we need to know about the idealized Otto engine cycle. The temperature at point 2 is determined by applying the following equation (valid for a reversible, adiabatic compression of an ideal gas with constant heat capacities) to step 1–2:

$$T_2 = T_1 \left(\frac{P_2}{P_1}\right)^{(\gamma-1)/\gamma} = (600)(15)^{0.286} = 1300°\text{R } (840°\text{F})$$

Calculation of T_4 is similar because step 3–4 is a reversible, adiabatic expansion. However, it is simpler to use the equation connecting T and V for such a process; thus we have

$$\frac{T_4}{T_3} = \left(\frac{V_3}{V_4}\right)^{\gamma-1}$$

But the analogous equation for step 1–2 is

$$\frac{T_1}{T_2} = \left(\frac{V_2}{V_1}\right)^{\gamma-1} = \left(\frac{V_3}{V_4}\right)^{\gamma-1}$$

The last equality comes about because $V_2 = V_3$ and $V_1 = V_4$. Thus, by comparing these two equations, we have

$$T_4 = T_3 \frac{T_1}{T_2} = (4400)(600/1300) = 2030°\text{R } (1570°\text{F})$$

The following list summarizes the temperatures at the various points of the cycle:

$$T_1 = 600°\text{R } (140°\text{F})$$
$$T_2 = 1300°\text{R } (840°\text{F})$$
$$T_3 = 4400°\text{R } (3940°\text{F})$$
$$T_4 = 2030°\text{R } (1570°\text{F})$$

These temperatures are all we need for calculation of the work and heat quantities associated with each step. For an ideal gas with constant C_V, the internal energy change is always given by $\Delta U = C_V \, \Delta T$. From the First Law,

$$\Delta U = Q - W$$

Therefore

$$C_V \, \Delta T = Q - W$$

Now either Q or W is zero for each step of the cycle. Two steps are adiabatic, and $Q = 0$; two steps are at constant volume, and $W = 0$. Thus we have

Step 1–2: $Q_{12} = 0$ $\quad W_{12} = -C_V \, \Delta T = -5(1300 - 600) = -3500$ Btu

Step 2–3: $W_{23} = 0$ $\quad Q_{23} = C_V \, \Delta T = 5(4400 - 1300) = 15{,}500$ Btu

Step 3–4: $Q_{34} = 0$ $\quad W_{34} = -C_V \, \Delta T = -5(2030 - 4400) = 11{,}850$ Btu

Step 4–1: $W_{41} = 0$ $\quad Q_{41} = C_V \, \Delta T = 5(600 - 2030) = -7150$ Btu

All of these results are shown on Fig. 3-2. The total work of the cycle is $W_{cycle} = -3500 + 11{,}850 = 8350$ Btu.

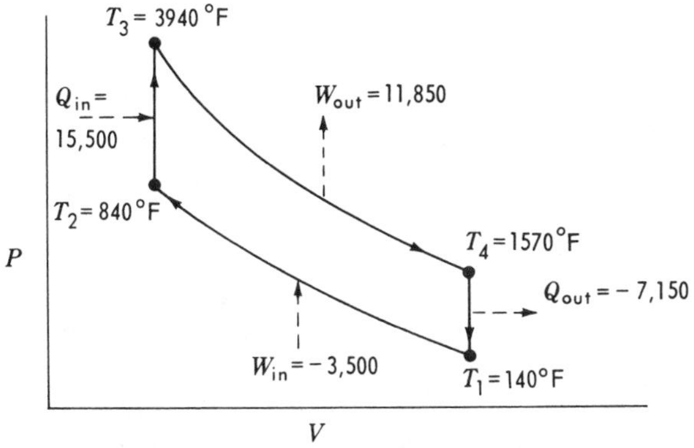

Figure 3-2 Results for idealized Otto cycle.

The thermal efficiency of the cycle η is defined as the ratio of the work of the cycle to the heat added; that is,

$$\eta = \frac{W_{cycle}}{Q_{in}} = \frac{W_{cycle}}{Q_{23}} = \frac{8350}{15{,}500} = 0.54$$

This means that only 54 percent of the heat added to the engine is converted into work. The remaining 46 percent leaves the engine as heat $Q_{out} = Q_{41} = -7150$ Btu. This result is obtained for an ideal cycle, considered to operate reversibly. We can think of no way to improve this result for the same conditions of operation.

Now an actual engine will not execute the idealized cycle. The compression and expansion steps will be neither reversible nor adiabatic. The work output of the engine will be considerably less than the work we have calculated for the idealized cycle. One way to estimate the work of a real engine that accomplishes the *same changes of state* as occurred in our ideal cycle is to assign an efficiency to each of the steps of the cycle. For steps that produce work we define our efficiency so that it gives the ratio of the actual work produced to the reversible work; that is,

$$\text{efficiency} = W_{\text{prod}}/W_{\text{rev}}$$

This efficiency is always less than unity, and it is a measure of the extent to which irreversibilities reduce the work output in comparison with the maximum possible work. For steps that require work, the reversible work is the minimum necessary, and irreversibilities can only serve to increase the work requirement. Thus we define our efficiency in this case as $W_{\text{rev}}/W_{\text{req}}$. Again the efficiency is always less than unity.

The only basis on which we can assign a numerical value for an efficiency is experience, either our own or that recorded by others. In the case of compression and expansion processes, experience shows that an efficiency of about 0.75 is a reasonable figure. Thus for step 1–2, where W_{12} is required, we can now estimate a reasonable value as follows:

$$W_{12} = \frac{W_{\text{rev}}}{0.75} = \frac{-3500}{0.75} = -4670 \text{ Btu}$$

From the First Law we have

$$C_V \, \Delta T = Q_{12} - W_{12}$$

or

$$Q_{12} = W_{12} + C_V \, \Delta T$$

$$Q_{12} = -4670 + (5)(1300 - 600) = -1170 \text{ Btu}$$

Thus we see that the actual irreversible process cannot be adiabatic, and 1170 Btu is transferred from the gas in the cylinder to the surroundings. Nor is this in any way unreasonable in view of the elevated temperatures of the gas and the fact that a real engine is cooled by either water or air.

The expansion step 3–4 produces work, and we calculate an estimate of the actual value as follows:

$$W_{34} = (0.75)W_{\text{rev}} = (0.75)(11{,}850) = 8900 \text{ Btu}$$

Again we calculate the heat transferred by the First Law; thus

$$Q_{34} = W_{34} + C_V \, \Delta T$$
$$Q_{34} = 8900 + (5)(2030 - 4400) = -2950 \text{ Btu}$$

The actual expansion is also seen to result in the transfer of heat from the gas in the cylinder to the surroundings. This heat transfer is even larger than that for step 1–2 because the gas temperatures are higher. Steps 1–2 and 3–4 are the only ones involving work, and the work of the engine is the sum of W_{12} and W_{34}; we then have

$$W_{\text{engine}} = -4670 + 8900 = 4230 \text{ Btu}$$

The thermal efficiency of the actual engine is again given by the net work W_{engine} divided by Q_{23}, the heat added to the engine (by burning fuel), as follows:

$$\eta = \frac{W_{\text{engine}}}{Q_{23}} = \frac{4230}{15{,}500} = 0.27$$

The irreversibilities which we have assumed to exist in our engine by taking an efficiency of 0.75 for each of the steps of the cycle have reduced the thermal efficiency η by a factor of 2, and a value of $\eta = 0.27$ for a real engine operating on the Otto cycle is a reasonable value for a well-tuned engine operating under conditions for which it was designed.

Whether you realize it or not, we have been talking about a *heat engine* (this is a particular kind of heat engine known as an internal-combustion engine, but it serves to introduce the general topic of heat engines). The topic is of importance since more than 90 percent of our power is supplied by heat engines, and no one expects heat engines to become any less important in the foreseeable future. There is great interest in the development of new *kinds* of heat engines, but the expectation is that most of our power will continue to come from some kind of heat engine. Nuclear power plants, magnetohydrodynamic generators, thermoelectric generators, thermionic generators, jet engines, and rockets are all heat engines and are subject to the same thermodynamic analysis as is the internal-combustion engine.

It all began about the year 1700. A couple of steam pumps were invented a few years earlier, but the first real steam *engine* was invented by Newcomen in 1712. It was a most inefficient contraption until James Watt, starting 50 years later, introduced a number of innovations which made the steam engine a practical device. Thus Watt did for the steam engine what Otto was to do 100 years later for the internal-combustion engine. Watt not only developed but also manufactured the steam engine, which became widely used. Fulton's steamboat, the Clermont, was driven by one of Watt's engines on its trips between New York and Albany starting in 1807.

In 1824 a young Frenchman named Carnot published a paper on the motive power of heat, the first theoretical consideration of heat engines. The steam engine was well known to Carnot. He knew that it had been made increasingly efficient over the years, and he wondered whether there was some limit to its improvement. He appreciated that real steam engines leaked steam and that friction reduced their efficiency. So he imagined the ideal engine, one that we would call reversible, then he formulated the problem in exactly the right way. This was his stroke of genius, for once having recognized the problem, he could hardly have failed to solve it.

Imagine that we build a fire under some object so that we may keep it at a constant temperature T_H well above the temperature of the surroundings at T_C. Thus both the object and the surroundings constitute *heat reservoirs* at constant temperature. Now we wish to operate a reversible heat engine in such a way that it takes in heat at the temperature T_H and *only* at T_H, and discards heat, if this is necessary, at T_C and *only* at T_C; the engine is to operate continuously. We must answer the following question: must heat be discarded at T_C, and if so, how much? We imagine our engine to consist of the familiar piston-and-cylinder arrangement containing a gas as the working medium. Since the engine is to perform continuously, the piston and cylinder with its contents must periodically return to their initial states, i.e., the engine must operate in a cycle. The statement of the problem imposes restraints on what kinds of steps the cycle may contain. We said that heat transfer may occur only at the temperatures T_H and T_C. Thus any steps during which heat may be transferred must be isothermal, either at T_H or at T_C, and any other step must then necessarily be adiabatic. The cycle must consist of steps which represent isothermal processes at T_H and T_C and of steps representing adiabatic processes between these two temperatures. The only possible such combination of steps representing a work-producing cycle is shown in Fig. 3-3*a*. Starting at point 1, where the piston is at the inner limit of its stroke, we have the apparatus in thermal equilibrium with the hot reservoir at T_H. The gas now expands reversibly and isothermally, doing work against the piston as it follows the path 1–2. During this step, heat Q_H must be added to the gas in order to keep the temperature constant at T_H. If heat were not added during this process, the temperature would drop because the work of expansion would be produced at the expense of the internal energy of the gas, and this decrease in internal energy would be reflected by a temperature decrease. This is exactly what happens in the next step of the cycle, step 2–3, where the gas expands reversibly and adiabatically. At point 3 the piston has reached the outer limit of its stoke and must begin its return. The path of the process could just retrace that already followed, but this would require exactly the work done in steps 1–2 and 2–3, and the engine would then produce no work. The only alternative is to follow the path 3–4–1. During step 3–4 the process is reversible and isothermal at temperature T_C. It is a compression process and requires work. This work input would raise the temperature of the gas if no heat were transferred to the cold heat reservoir at T_C. Thus the heat Q_C must be extracted to make the process isothermal. At point 4 the heat transfer is interrupted, and the remaining compression step 4–1 occurs reversibly and adiabatically. The cycle of steps encloses the shaded area of Fig. 3-3*a*, and this area represents the $\int P\,dV$

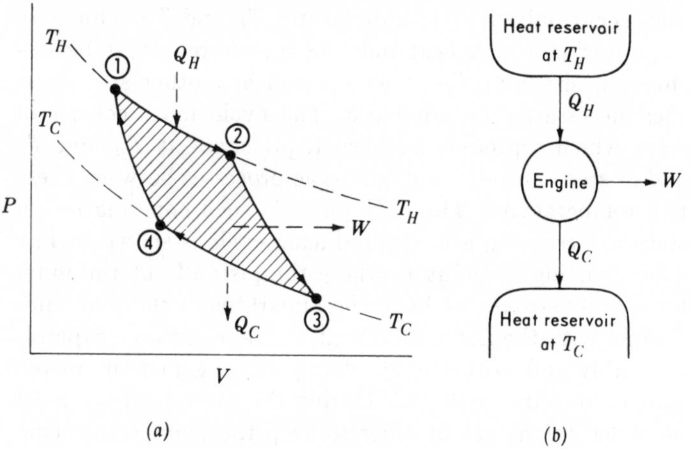

Figure 3-3 The Carnot engine cycle.

taken around the entire cycle; this is the net work of the engine as it executes one complete cycle.

Figure 3-3*b* is a schematic diagram which shows the various energy exchanges accomplished by the Carnot engine. The First Law applied to the engine shows that

$$\Delta U = Q_H + Q_C - W$$

where ΔU is the internal energy change of the engine itself.

However, for a complete cycle, ΔU must be zero because the engine returns to its initial state. Thus

$$W = Q_H + Q_C$$

and the thermal efficiency of the Carnot engine is given by

$$\eta_{\text{Carnot}} = \frac{W}{Q_H} = \frac{Q_H + Q_C}{Q_H} = 1 + \frac{Q_C}{Q_H}$$

Since Q_C represents heat leaving the engine, it has a negative value, and η is seen to be less than unity.

The only physical parameters specified for the Carnot engine cycle were the temperatures T_H and T_C. Therefore, it seems clear that the ratio Q_C/Q_H can depend only on T_H and T_C. The nature of this dependence turns out to be very simple and is provided by the equation

$$\frac{-Q_C}{Q_H} = \frac{T_C}{T_H} \tag{3-1}$$

This is proved in virtually every thermodynamics textbook ever written, and the proof will not be repeated here. The consequence is that we obtain the following very simple expression for the thermal efficiency of a Carnot engine:

$$\eta_{\text{Carnot}} = 1 - \frac{T_C}{T_H} \tag{3-2}$$

We can now answer our initial question (must heat be discarded at T_C, and if so, how much?). Equation (3-1) shows that

$$Q_C = -\frac{T_C}{T_H} Q_H$$

Thus, for given values of Q_H and T_H, Q_C depends only on the temperature of the cold reservoir T_C, and this is limited by the temperatures naturally available to us. These temperatures are well above absolute zero, and we have no practical means by which to reduce Q_C to a negligible value. This means also, as seen from Eq. (3-2), that the thermal efficiency of a Carnot engine, operating between temperatures that can be realized in practice, cannot approach unity. This conclusion, reached through a study of Carnot engines, applies equally to heat engines of all kinds. The Otto engine, considered earlier, is certainly no exception, for the efficiencies we calculated were well below unity. Thus all heat engines convert only a part of their heat intake into work and discard the remainder to the surroundings. This limitation on heat engines is not contained within the First Law of Thermodynamics. Nor does it result from imperfections in the engines, for we have found it by examining processes carried out reversibly, that is, as perfectly as we can imagine. This suggests that there must be a Second Law of Thermodynamics which imposes limits not expressed by the First Law. There is indeed a Second Law, and we will consider it in detail in a later chapter.

4

Power Plants

In the last chapter I brought out some of the fundamental properties of heat engines. Now let's see how these may be used to make some very elementary calculations with respect to a stationary power plant that generates electricity by expanding steam through a turbine. The basic scheme for operation of such a plant is shown in Fig. 4-1. There are four primary devices in the power cycle. The boiler serves to convert liquid water into steam at a high pressure and a high temperature. This requires heat from some high-temperature source; this heat is shown as Q_H. The steam so generated is fed to a turbine which drives an electric generator. This turbogenerator set is, of course, the heart of the power plant. Steam expands through the turbine and exhausts at a low pressure. This expansion process occurs adiabatically and is as nearly reversible as possible, thus providing the motive power to drive the turbine. The exhaust from the turbine enters a condenser, which is water-cooled, and this causes the steam to condense and the heat Q_C to be removed. The liquid condensate is then pumped into the boiler where it is revaporized. A small fraction of the work of the turbine is required to operate the pump.

The heat Q_H required by the plant may be supplied either by burning a fossil fuel, such as coal or oil, or it may come from a nuclear reactor. It makes little difference to the operation of the steam cycle shown in Fig. 4-1. This cycle

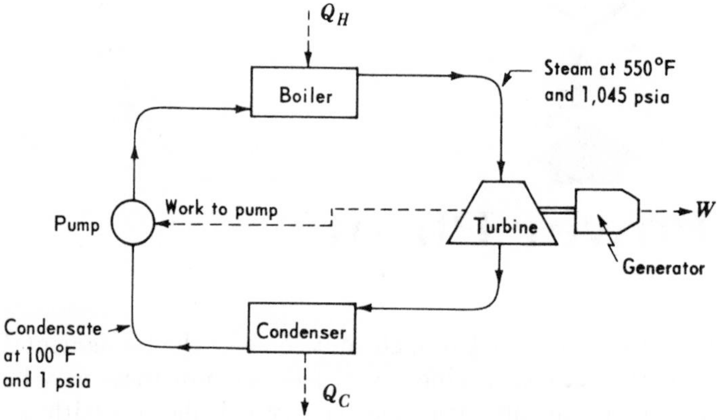

Figure 4-1 Steam cycle for stationary power plant.

departs only slightly from that of a Carnot engine. The boiler and condenser accomplish changes which are largely isothermal at the boiler temperature T_H and at the condenser temperature T_C. The processes within the turbine and pump are as nearly adiabatic and reversible as they can be made.

One may wonder why the steam from the turbine is not

simply discarded to the atmosphere. Fresh water would then be fed continuously to the boiler. There are several reasons why this is not done. In the first place, it would mean that atmospheric pressure would be the lowest pressure to which steam could be expanded. When a condenser is used, the system is closed, and the condenser pressure (at which the turbine exhausts) can be maintained at values well below atmospheric. This extra expansion of the steam allows the turbine to do more work. Equally important is the fact that the water used in the steam cycle must be very pure. The reason is that all dissolved solids are left behind in the boiler when the water is vaporized. Any appreciable accumulation of such solids would foul the heat-transfer surfaces of the boiler and would eventually plug the boiler tubes. Thus any water added to the system must be carefully purified, and this is an expensive process. As a practical matter it is necessary to use the purified water over and over. Finally, the steam circulation rate, as we shall see shortly, is by no means small, and to discard the spent steam to the atmosphere could severely affect the humidity for miles around. Of course, condensation of the steam in the condenser means that a large quantity of heat Q_C must be discarded to the surroundings, usually to a river; there are problems with this, too, and I'll return to them later.

Let's be specific and consider a modern nuclear power plant typical of about 50 such plants now in various stages of design and construction in this country. The typical plant has a rated capacity of some 750,000 kw, or about a million horsepower. Steam is generated in the boiler at about 550°F and 1,045 pounds per square inch absolute pressure (psia). The condenser operates at about 100°F and 1 psia. These values are shown on Fig. 4-1. Thus the heat source for operation of the plant, that is, the nuclear reactor which supplies Q_H, must be maintained at a temperature T_H of at least 550°F (or 1010°R), and the heat sink, that is, the surroundings which absorb the heat Q_C rejected by the condenser, must have a temperature T_C of no more than 100°F (or 560°R). For a Carnot engine operating between these two temperatures the thermal efficiency would be

$$\eta = \frac{W}{Q_H} = 1 - \frac{T_C}{T_H} = 1 - {}^{560}\!/_{1010} = 0.445$$

Thus at best our power plant could convert only 44.5 percent of the heat taken in from the nuclear reactor Q_H into work, and the other 55.5 percent would have to be discarded to the surroundings as Q_C. No actual plant can possibly be as efficient as a Carnot engine, which presupposes perfection. So we will take $\eta = 0.30$ as a realistic value for the thermal efficiency of our plant. Since by definition

$$\eta = 0.30 = \frac{W}{Q_H}$$

we have

$$Q_H = \frac{W}{0.30} = \left(\frac{750{,}000 \text{ kw}}{0.30}\right)\left(57 \frac{\text{Btu/min}}{\text{kw}}\right) = 1.42 \times 10^8 \text{ Btu/min}$$

and

$$Q_C = (0.7)Q_H = (0.7)(1.42 \times 10^8) = 1.0 \times 10^8 \text{ Btu/min}$$

The difference between these two heat rates is the rate of production of work, 0.42×10^8 Btu/min, which corresponds to a power output of 750,000 kw.

Now from data for the heat of vaporization of water, we know that about 1100 Btu is required for each pound of water entering the boiler and vaporized there. Thus the steam circulation rate is

$$\dot{m} = \frac{1.42 \times 10^8 \text{ Btu/min}}{1100 \text{ Btu/lb}} = 1.3 \times 10^5$$

or 130,000 lb/min or 65 tons/min. This confirms my earlier statement that the steam rate is no small number. You may think that this is a heat engine on too grand a scale, but I reiterate my earlier statement that about 50 such nuclear plants are now under construction or design. Moreover, about 50 additional plants depending on fossil fuels are also under construction or design. Each such plant costs upwards of $100 million. Moreover, these plants are designed around a single turbogenerator; one unit produces 1,000,000 hp. Such a turbogenerator set weighs more than 2,000 tons and occupies a space on the order of two-thirds of a football field.

How big must the pipe be that supplies steam to such a turbine? For steam at 550°F and 1,045 psia, the specific volume is about 0.42 ft³/lb, and a reasonable velocity for high-pressure steam in a pipe is 75 ft/sec. The volumetric flow rate of steam is given both as the product of the velocity u and the cross-sectional area A of the pipe and as the product of the specific volume V of the steam and its mass flow rate $\dot{m}$. Thus

$$uA = V\dot{m} \qquad \text{or} \qquad A = \frac{V\dot{m}}{u}$$

and

$$A = \frac{(0.42\ \text{ft}^3/\text{lb})(130{,}000/60\ \text{lb/sec})}{75\ \text{ft/sec}} = 12.1\ \text{ft}^2$$

A pipe 4 ft in diameter has just a bit more than this cross-sectional area. A more suitable choice would be a design with four pipes 2 ft in diameter. If this seems large, consider the exhaust duct of the turbine, which carries steam that has expanded to a specific volume of about 300 ft³/lb. Higher velocities are allowed here, and we take a value of 500 ft/sec to be reasonable. Thus

$$A = \frac{V\dot{m}}{u} = \frac{(300\ \text{ft}^3/\text{lb})(130{,}000/60\ \text{lb/sec})}{500\ \text{ft/sec}} = 1{,}300\ \text{ft}^2$$

This requires a duct some 36 ft square to connect the turbine with the condenser.

We can calculate still other quantities. For a nuclear power plant the heat Q_H comes from the reactor core at the expense of mass, according to the formula

$$Q_H = \frac{-c^2\,\Delta m}{g_c}$$

where g_c is a dimensional constant equal to 32.17(lb mass/lb force)(ft/sec^2). Thus

$$\Delta m = \frac{-Q_H g_c}{c^2}$$

$$= \frac{\left(-1.42 \times 10^8 \frac{\text{Btu}}{\text{min}}\right)\left(778 \frac{\text{ft-lb force}}{\text{Btu}}\right)\left(32.17 \frac{\text{lb mass-ft}}{\text{lb force-sec}^2}\right)}{10^{18}\ \text{ft}^2/\text{sec}^2}$$

$$\Delta m = -3.5 \times 10^{-6}\ \text{lb mass/min or about } -2\ \text{lb/year}$$

This is a figure one would need before designing the nuclear reactor, along with a lot of additional information.

The heat discarded by the condenser, Q_C, must go somewhere and is almost always dumped into a river. No small river will do, as the amount of heat involved is enormous, on the order of 10^8 or 100,000,000 Btu/min. Consider, for example, the upper Hudson river, which has been considered as the site for a nuclear plant. The average flow rate of the Hudson is about 5,800 ft^3/sec or about 0.22×10^8 lb/min. Each pound of river water will be raised in temperature 1°F for each Btu of heat it adsorbs. Thus in absorbing 1.0×10^8 Btu/min the 0.22×10^8 lb/min of river water will rise in temperature by

$$\Delta T = \frac{1.0 \times 10^8}{0.22 \times 10^8} = 4.5°\text{F}$$

The flow rate of the Hudson River is controlled, and if the minimum flow rate is held above 3,600 ft^3/sec, the maximum temperature rise of the river would be about 7°F. Of course, not all of the river is diverted through the condenser of the power plant, but the temperature rise of the water flowing through the condenser is limited because the water temperature cannot exceed the condensation temperature of the steam, about 100°F. If about 20 percent of the river is diverted to run through the condenser, the temperature rise of this portion of the river is

$$\Delta T = \frac{1.0 \times 10^8}{(0.22 \times 10^8)(0.2)} = 22.5°\text{F}$$

For river temperatures below 70°F, this temperature rise is quite acceptable. When returned to the river, this hotter stream is diluted by the undiverted portion of the river to produce the overall temperature increases of the river already calculated. The river water used to absorb the discarded heat of the plant must be pumped through the condenser, and this requires a considerable expenditure of work. In fact, the work required to run the plant itself amounts to something like 50,000 hp.

If this power plant were coal-fired rather than a nuclear plant, the quantities we have calculated would be little different. We would have essentially the same amount of heat to discard, and the temperature rise or "thermal pollution" of the river would be about the same. But there are other considerations. To supply the heat Q_H with coal having a heating value of 12,000 Btu/lb (a usual value) would require the burning of coal in the amount of

$$\frac{1.42 \times 10^8 \text{ Btu/min}}{12{,}000 \text{ Btu/lb}} = 12{,}000 \text{ lb/min}$$

or about 6 tons/min or 3,100,000 tons/year, and this assumes that the entire heating value of the coal can be used, which it cannot. The sulfur content of coal used to fire power plants averages about 2½ percent, and the combustion of this coal at the required rate would produce some 600 lb of sulfur dioxide per minute. This noxious gas would enter and pollute the atmosphere continuously. Thus the coal-fired plant would cause both thermal pollution and air pollution.

This brings me to a very important matter. What are the consequences of building and operating a nuclear power plant such as I have described? I should remark at once that it is the *purpose* of engineering to build and operate such plants—to build and operate many different kinds of plants, all of them having an end product which is in some sense useful to mankind. Thus one consequence of building a power plant is the continued supply of ample power at a reasonable cost. Other less desirable consequences come as by-products. Pollution is one of these by-products, and it was until recently largely ignored. Thirty years ago hardly anyone thought that industrial plants, built to produce the things people need, want, or think they want, could possibly have as a side effect the alteration or pollution of the environment to a significant and serious extent. But the rapid growth of both population and affluence has now made this a major concern, and no engineer can ignore it.

Yet it presents a dilemma, which is illustrated by our consideration of power plants. The utility companies would hardly be constructing billions of dollars worth of power plants without the conviction that the public will *demand* the output of these plants. Even the most ardent conservationist would be indignant if told his electric power consumption would be rationed because the utility companies had decided not to increase generating capacity because of the danger of greater pollution. Actually, nuclear power plants do not contribute to air pollution, but there seems to be no way to prevent thermal pollution. You may say that one more power plant on a given river that raises its temperature a mere 4 or 5°F can hardly be a serious problem, and you are probably right. But where in 5 years do we build the next plant, and the next, and the next? We see no early voluntary end either to increasing population or to greater demand for more and more of the products of engineering technology. Various governmental agencies have embarked on extensive and expensive plans to clean up our rivers. Is this being done so that we can turn them into steaming heat sewers in which and along which nothing but algae can grow? I am not suggesting that any single plant presents a clear and present danger. I am pointing to a problem that will become critical in the foreseeable future. It is just one of the extraordinarily difficult problems brought into being by our expanding industrial society and one which will be left to the next generation to solve. It's a problem that owes its existence to the Second Law of Thermodynamics, to which I will direct attention in subsequent chapters.

5

The Second Law of Thermodynamics

It is essential at this point that we consider the Second Law of Thermodynamics. If you have already studied this subject from the traditional point of view, no harm is done. Our perspective departs somewhat from the traditional, but the conclusions reached are the same.

We can ask right at the start why we should be looking for a Second Law at all, and this takes a little explaining. It is a matter of everyday experience that certain processes *do* and certain processes *don't* occur in the world. It's very easy to give simple examples. For illustration, I will consider two types of processes. The first is strictly mechanical in nature and makes use of our old standby, the piston and cylinder. We will consider adiabatic processes carried out on a gas in this apparatus.

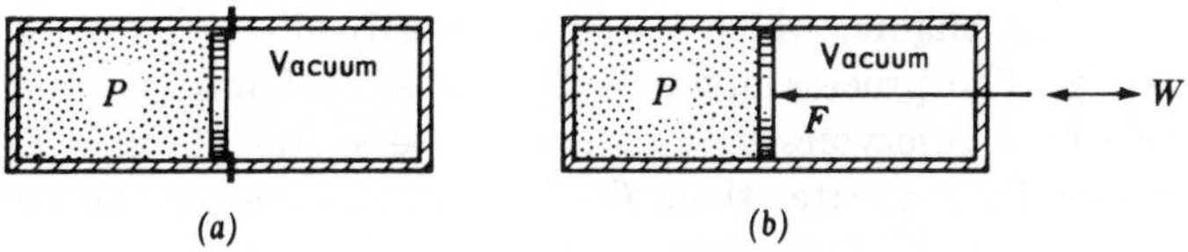

Figure 5-1 Adiabatic mechanical process.

Figure 5.1*a* shows a gas confined by a piston held by latches. If the latches are removed, we have no doubts about what will happen. We need no law to tell us that the piston will move right and not left. Of course, such a process does nothing for us, but we readily see that we can put it to use by opposing the motion of the piston by a force on its right side (see Fig. 5.1*b*). Now if the force is very small, we will get very little work. As the force is increased, we get more and more work. But there is a limit, and this comes when we make the force equal to PA. Now we have a balance of forces, and we know that nothing will happen. But if this force is reduced differentially, we get the maximum possible work out. We have already characterized this process as reversible, and we know that $W_{rev} = \int P\,dV$. On the other hand, if the force is made differentially larger than PA, then the piston moves to the left, and work is put into the system. Again we have a reversible process with W_{rev} given by $\int P\,dV$. If the force is increased further, latches are required on the opposite side of the piston to hold it, and again we know the direction the process will take if the latches are removed. In this case the work done on the gas will be *greater than* the reversible work. We see a whole gradation of processes, and we see the uniqueness of the reversible process. It represents an upper limit for the expansion process which produces work and a lower limit for the compression process which requires work, and thus it serves as a point of departure for discussion of processes in either direction.

The second type of process is thermal and makes use of the heat engine, which we have already discussed extensively. The process involves the transfer of heat between two heat reservoirs held at the temperatures T_H and T_C, where T_H is greater than T_C. In Fig. 5.2*a* we see the two

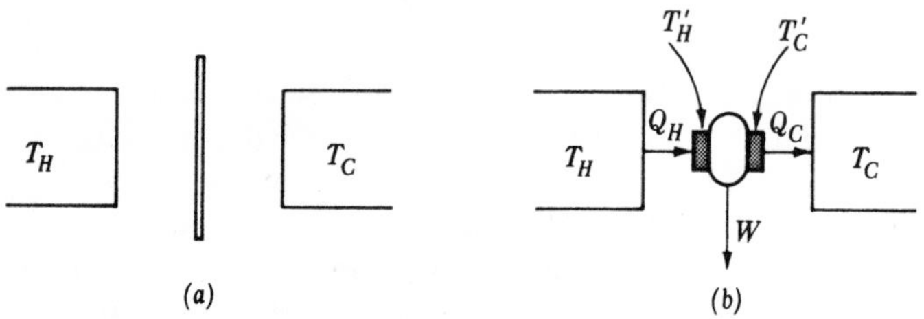

Figure 5-2 Thermal process.

heat reservoirs separated by a perfect insulator. If the insulator is removed, we have no doubts about the direction of the heat flow. Again, such a process does no work, but we can make use of it by interposing a heat engine between the two reservoirs. Now the heat engine must have a part to which heat is added and a part from which heat is discarded, and we will assume these are maintained at the temperatures T'_H and T'_C (see Fig. 5.2*b*). For a stationary power plant these parts would be the boiler and the condenser. If the engine itself is reversible, then

$$W = Q_H\left(1 - \frac{T'_C}{T'_H}\right)$$

and the amount of work done will depend on the temperature differences $T_H - T'_H$ and $T'_C - T_C$, or equivalently on the difference $T'_H - T'_C$. If $T'_H = T'_C$, then no work is done, and the process occurs just as if the engine were not there.

As $T'_H - T'_C$ becomes larger, more and more work is done. However, there is a limit, and this is reached when T'_H and T'_C become almost equal to the reservoir temperatures T_H and T_C. When this happens the process is reversible, and we realize the maximum amount of work. At this point we may reverse the entire process, putting work into the process and transferring heat in the opposite direction. This requires only that T'_C be less than T_C and that T'_H be greater than T_H by differential amounts. We then have a reversible *heat pump* or refrigerator, one that requires the minimum work for a given refrigeration effect. If we go further and make $T'_C < T_C$ and $T'_H > T_H$ by finite amounts, then the process becomes irreversible again, and the required work increases. Again we see a whole gradation of processes, with the reversible process appearing as unique.

These two types of processes, one mechanical and the other thermal, are clearly different. Yet the intriguing thing is that we can say the same kinds of things about both. In each case we readily see the *cause* of the process and can correctly predict the changes that will occur—the "direction" of each process. In each case we see how to make the process reversible so that it produces the maximum work. We also see how to carry out the reverse process, both reversibly and irreversibly. The reason everything is so obvious is that each process is a part of our experience. As long as we talk about simple processes that are readily connected to our previous experience, we can rely on that experience to predict what will happen. Difficulties arise only when we encounter more complex situations which have no obvious connection with our past observations.

For example, a man might walk in here with a box 2 ft long and 6 in. on a side and say, "Look here, I've got a box of tricks! See this pipe sticking out the top and the pipes sticking out either end? If you hook a compressed-air line to the center pipe, cold air will come out the left end and hot air will come out the right end. How about that?"

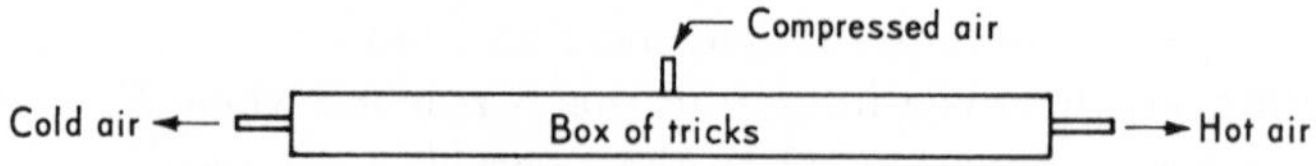

Figure 5-3

Can it work, or can't it work? Unless you have seen it before, you can only guess, and you will most likely base your guess on how you size up the man. But there must be a way to tell whether or not this process can or cannot take the "direction" claimed by the man. Somehow we need to generalize the results of our previous experience with the simple processes we already know about. And this is the problem. How do we generalize in a meaningful way when the things we have observed are all so different in character from one another? We might look for some easily understood statement in common words that provides an obvious answer to the question, "Is it or isn't it possible?" But what statement should we make? A number have been proposed, and I'll give you a few so that you can judge how directly they relate to the question of whether the box of tricks the man brought in can do what he says it will:

1. "No engine, operating in a cycle, can convert all the heat it takes in into work." Now what connection does the man's device have with an engine? He says, in fact, that it has no moving parts. Moreover, he says it is thoroughly insulated; thus there is no heat taken in.
2. "Heat cannot be caused to flow from a cooler to a hotter body without producing some other effect." Again, there is no heat transfer discernible in this box of tricks, and we see no obvious connection between this statement and the accomplishments of the device.
3. "Processes occur in a direction so as to progress from nonequilibrium toward equilibrium states." Now who is to say that two air streams at different temperatures but both at atmospheric pressure represent something

closer to equilibrium than one stream of compressed air at a different temperature?

One can surely make a case that each of these statements is true. But on which one would you like to stake your claim that the man with the box of tricks is either a fraud or a genius? It is very hard to see the relevance of any of these statements. Moreover, all are qualitative. How could we use such statements to allow us to make quantitative calculations? You must be wondering why, if such statements are so useless, they form the starting points for the traditional treatment of the Second Law. I wonder why, too, and I'll return to the question later. For the moment we will forget them and look for another means (one that is quantitative) by which to make a generalization of our experience.

The fact that we are looking for something quantitative suggests that we ought to be looking for a *property*. But this property can hardly be obvious, or we would already be talking about it. It must be a property (like internal energy) that cannot be directly observed but which is a function of other variables. How are we to discover such a property? The only way is to carry out a series of controlled experiments, making all the measurements we can. Then we examine the data to see if any consistent pattern emerges to suggest the existence of a new property.

Thus I ask you to imagine another set of experiments done with our reliable old piston-and-cylinder assembly. Again, for ease of visualization we take the material contained in the cylinder to be a gas—any old gas; it need not be ideal. Imagine that we do a whole series of experiments in which we always start with a fixed amount of gas in our cylinder at fixed conditions of temperature and pressure, say, T_1 and P_1, and in every experiment we change these conditions to the same final values T_2 and P_2. But each experiment is different from the others in that between the initial state 1 and the final state 2 the relation between T and P, that is, the *path*, is different from all the others. We can easily alter the path from experiment to experiment by changing the amounts and relative rates of heat and work addition or extraction.

It turns out that the only processes for which we can really make precise and detailed measurements are those carried out very slowly, those which approximate reversibility. Any process carried out rapidly will cause nonuniformities of temperature and pressure within the gas, and for such a process we cannot know the exact path, or the precise P-T relation, for we don't know what P and what T to talk about. Thus we restrict our experiments to those carried out reversibly, or almost reversibly. Such processes are unique, and as we have seen they provide a standard against which to measure all other processes and a convenient point of departure for the consideration of all other processes.

Thus we carry out various reversible processes by which we proceed from state 1 of the gas to state 2, and we keep a careful accounting of all measurable variables from point to point in each experiment. The apparatus is shown in Fig.

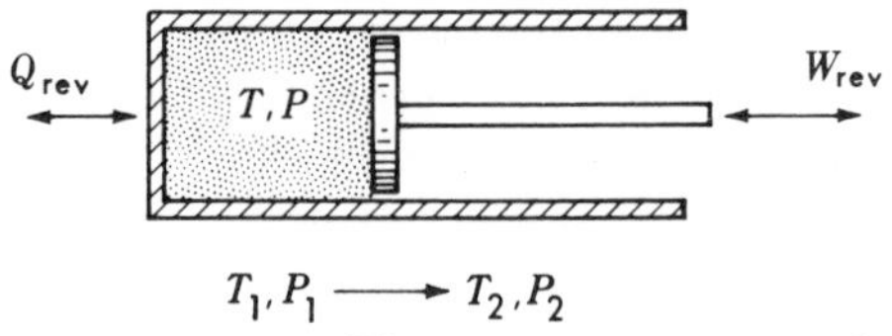

Figure 5-4

5-4, and a graph indicating several possible P-vs.-T relations or paths appears in Fig. 5-5. Each such path represents one experiment, and for each experiment we may make a table, such as Table 5-1, which lists the data taken as we progress from point 1 to point 2 along any one path shown in Fig. 5-5.

*Table 5-1**

T	P	V	Q_{rev}	W_{rev}
T_1	P_1	V_1	0	0
↓	↓	↓	↓	↓
T_2	P_2	V_2	Q^f_{rev}	W^f_{rev}

* The superscript f to Q_{rev} and W_{rev} indicates a final value.

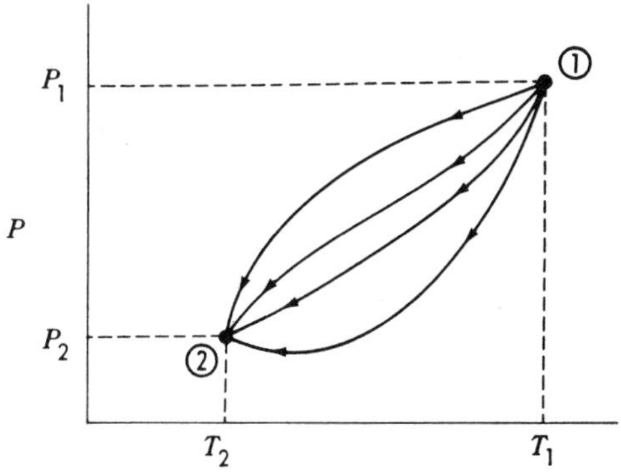

Figure 5-5

Eventually we prepare many such tables of data, one for each of the processes by which we get from state 1 to state 2. Having this mass of data, we have a look to see whether any regularities occur from one set to the next. It doesn't take us long to notice that V_2 and V_1 are the same for all processes and thus that ΔV is the same for all. This is hardly a surprise, and it merely confirms that V is a property of the system and is a function of T and P. Or to put it more elegantly, an equation of state exists connecting P, V, and T, something we might have assumed from the beginning. The fact that we find it to be so does give us confidence in our experimental methods. Next we notice that although Q^f_{rev} has different values for all the processes and although W^f_{rev} has different values for all the processes, the difference $Q^f_{rev} - W^f_{rev}$ is the same for all processes. If this is so, then this difference must be the measure of a property change between state 1 and state 2. Indeed it is; the property change is $\Delta U = U_2 - U_1$, the change in internal energy, and we have rediscovered the First Law of Thermodynamics, that is, $\Delta U = Q^f_{rev} - W^f_{rev}$, for the special case of reversible processes connecting two fixed states. Again, we learn nothing new, except that our methods do produce results known to be right.

Thus far we've only looked at easy things, and we've a long way to go before we exhaust the possibilities. We can plot T vs. P, T vs. V, P vs. V, T vs. Q_{rev}, T vs. W_{rev}, P vs. Q_{rev}, and P vs. W_{rev}. Then we examine the graphs for some clue to the existence of a property. Perhaps some particular plot gives curves which all have the same area below them. However, as a matter of fact, we find nothing of the sort; so we get more subtle or more desperate and we start plotting reciprocals, and after a while we find something. Plots of $1/P$ vs. W_{rev} all have the same area below the curves! Here at last is what we've been looking for.

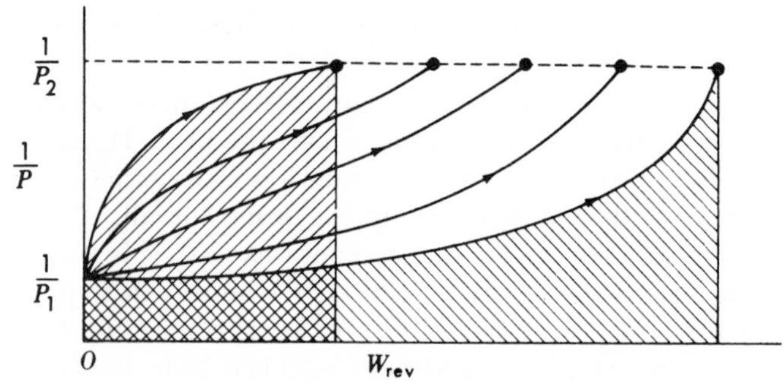

Figure 5-6

If all the areas are the same, this means that $\int_0^{W_{f\,\mathrm{rev}}} (1/P)\, dW_{\mathrm{rev}}$ is the same for all processes. Now, whenever we find a quantity that is the same for all processes between two fixed states, we know the quantity is independent of the path and must depend only on the initial and final states of the system. Thus the quantity must measure a *property change* of the system. All that remains is to give the property a symbol and a name. We will use the symbol X and name our property the Xtropy function. Now we can write

$$\int \frac{1}{P}\, dW_{\mathrm{rev}} = \Delta X$$

and we're pretty pleased with ourselves. But wait a minute. Now we notice that our value of ΔX is the same as ΔV. Is our Xtropy just the volume? It's easy enough to find out. Substitute ΔV for ΔX in our equation and see if it reduces to a known result; thus

$$\int \frac{1}{P}\, dW_{\mathrm{rev}} = \Delta V$$

or in differential form

$$\frac{dW_{\mathrm{rev}}}{P} = dV \qquad \text{or} \qquad dW_{\mathrm{rev}} = P\, dV$$

something we knew all along. So we've just rediscovered the volume, and all our work was for nothing. Well, not quite. We've actually seen something very important. It is that we *could* have discovered the *existence* of the property we call *volume* in this way. But we would not be able to give an explicit definition of it. All we would have is an equation, $dV = dW_{\mathrm{rev}}/P$, which tells us how to measure changes in this property. And if we were not able to measure volume directly and had to get along with just this equation, we would probably come up with some profound law, perhaps the law of volume conservation. This is exactly what happens with internal energy. We cannot measure or define it. We have merely an equation that tells us how to measure its change, $dU = dQ - dW$, and we have the law of energy conservation.

So it's back to the drawing board, but we do get an idea from all this. Since T is related to Q in a way quite similar to the relation between P and W, we guess that it might be interesting to plot $1/T$ vs. Q_{rev} for the various experiments we have made. So we do it, and sure enough the areas below these curves are all the same. Again we conclude that the quantity

$$\int_0^{Q_{f\,rev}} \frac{1}{T}\, dQ_{rev}$$

must be the measure of a property change. We decide to give the property the symbol N and to call it the *Ntropy function*, or more simply the *N*tropy. Thus we write

$$\int \frac{1}{T}\, dQ_{rev} = \Delta N$$

By now we are pretty wary, and we think perhaps we have again rediscovered something we already know all about. But this time we have not. This property *N*tropy is really new, and we know nothing more about it than what we have just found. We have no definition for it; we have only an equation telling us how to calculate changes. But we do know one very important thing: it is a property of the material in our cylinder and is a function of the conditions T and P. We know exactly as much about it as we do about internal energy. The only difference (and this may be the reason that energy seems less abstract) is that energy has several forms. We have not only internal energy but also potential and kinetic energy; however, there is no potential or kinetic *N*tropy, just internal *N*tropy.

Now it happens that symbolism is a very delicate matter among thermodynamicists. When they can't argue about anything else, they argue about that. And the symbol N we have taken for our new property turns out to be unacceptable. Forced to find another symbol, they agreed on S, perhaps the only complete agreement in the entire subject. But they kept the name, adding an initial "e" for the sake of euphony. Thus we have S = entropy, and

$$\int \frac{1}{T}\, dQ_{\text{rev}} = \Delta S$$

or in differential form

$$\frac{dQ_{\text{rev}}}{T} = dS \qquad \text{or} \qquad dQ_{\text{rev}} = T\, dS$$

Thus we have a new property, but we have not yet determined whether it has anything to do with the "directions" of processes. That was, after all, what we wanted a new property for. Let's return to the two processes we considered earlier. The first was a mechanical process—adiabatic expansion and compression. Consider first an adiabatic, reversible expansion. Since the process is reversible, we may use the equation

$$\Delta S = \int \frac{dQ_{\text{rev}}}{T}$$

and since it is adiabatic, dQ_{rev} is zero. Thus $\Delta S = 0$, and the process occurs at constant entropy; it is *isentropic*. We may represent the path of the process by a curve on a graph of P vs. T. It is shown on Fig. 5-7 as the solid curve drawn from point 1 to point 2. How would the process be changed were we to start again at point 1, but this time carry out an irreversible but adiabatic expansion to the same final pressure? We cannot really show the path of an irreversible process on our graph because there is no single P and no single T which apply to the whole system at any point of the process. So I have shown a series of hatch marks to represent a sort of average course for the process. It is, however, only the final state 3 reached by the system that

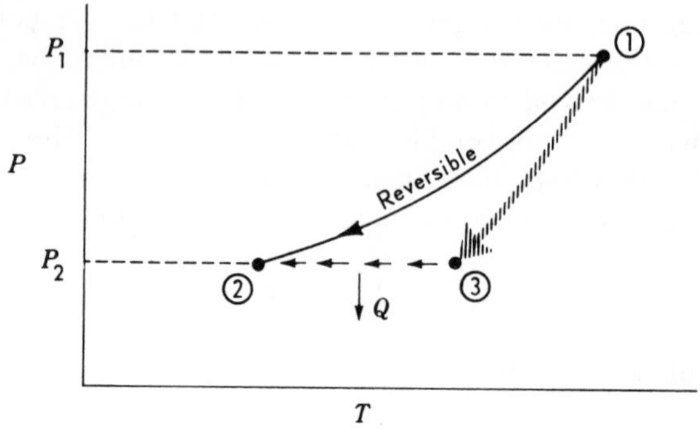

Figure 5-7 Adiabatic expansion.

really interests us. Since the process 1–3 is irreversible, it must produce less work than the reversible process 1–2. For an adiabatic process $\Delta U = -W$. Thus the irreversible process causes a smaller decrease in the internal energy of the gas than does the reversible process and consequently a smaller temperature decrease. From this we see that $T_3 > T_2$. Thus we could change from state 3 to state 2 by removing heat reversibly to decrease the temperature, and the resulting entropy change would be

$$\Delta S = S_2 - S_3 = \int \frac{dQ_{\text{rev}}}{T}$$

Since dQ_{rev} would be negative (heat removed), $S_3 > S_2$. On the other hand, the reversible process occurs at constant entropy; thus $S_2 = S_1$, and therefore $S_3 > S_1$. These results are summarized as follows:

Reversible process, 1–2	*Irreversible process, 1–3*
$S_2 = S_1$	$S_3 > S_1$
$\Delta S_{\text{total}} = 0$	$\Delta S_{\text{total}} > 0$

It should be noted that both processes 1–2 and 1–3 cause no entropy changes in the surroundings because both are adiabatic. Thus the entropy change of the system is the total entropy change.

Consider now the reverse process, that is, an adiabatic compression. This is shown on Fig. 5-8, where the solid

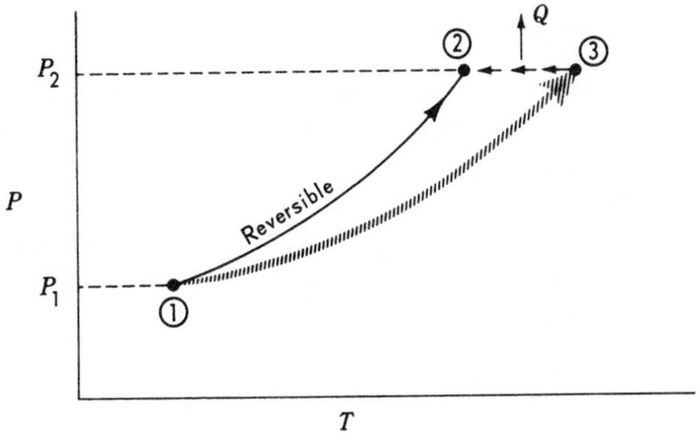

Figure 5-8 Adiabatic compression.

curve from point 1 to point 2 represents the reversible adiabatic process, and the hatch lines from 1 to 3 again approximate the irreversible adiabatic process. For these processes work is required, and the irreversible process requires more work than the reversible. Thus $\Delta U = -W$ will be positive (since work added is negative) and ΔU will represent a larger increase in internal energy for the irreversible process than for the reversible process. This means that $T_3 > T_2$, and again we could change from state 3 to state 2 by simply removing heat reversibly. Again,

$$\Delta S = S_2 - S_3 = \int \frac{dQ_{\text{rev}}}{T}$$

and it is negative because dQ_{rev} is negative (heat removed). So we have $S_3 > S_2$. But $S_2 = S_1$ for the reversible process, and therefore $S_3 > S_1$. These results are summarized as follows:

Reversible process, 1–2	*Irreversible process, 1–3*
$S_2 = S_1$	$S_3 > S_1$
$\Delta S_{\text{total}} = 0$	$\Delta S_{\text{total}} > 0$

Comparison with the results for the expansion process shows that with respect to entropy changes the same equations apply to both processes.

The other type of process I described was completely different in character and was centered around heat engines and heat pumps. For a heat engine operating reversibly we have the relation (see Chap. 3)

$$\frac{-Q_H}{Q_C} = \frac{T_H}{T_C} \qquad \text{or} \qquad \frac{Q_H}{T_H} + \frac{Q_C}{T_C} = 0$$

But Q_H/T_H is just ΔS_H, the entropy change of the heat reservoir at T_H, and Q_C/T_C is ΔS_C, the entropy change of the heat reservoir at T_C. Thus we have

$$\Delta S_H + \Delta S_C = \Delta S_{\text{total}} = 0$$

Since the engine operates in cycles and returns always to its initial state, there is no entropy change of the engine.

If the engine operates irreversibly, then it produces less work than the reversible work and must therefore discard a larger Q_C to the cold reservoir for the same heat intake Q_H. Thus for irreversible operation

$$\frac{Q_H}{T_H} + \frac{Q_C}{T_C} > 0$$

because Q_C is now larger than in the reversible case, and all other quantities remain the same. Thus for a heat engine operating irreversibly

$$\Delta S_H + \Delta S_C = \Delta S_{\text{total}} > 0$$

In summary:

Reversible heat engine	*Irreversible heat engine*
$\Delta S_{\text{total}} = 0$	$\Delta S_{\text{total}} > 0$

For the reverse process, the heat pump or refrigerator, work is done *on* the system; heat Q_C flows *out* of the heat reservoir at T_C, and heat Q_H flows *into* the heat reservoir at T_H. For reversible operation we have again

$$\frac{Q_H}{T_H} + \frac{Q_C}{T_C} = \Delta S_H + \Delta S_C = \Delta S_{\text{total}} = 0$$

The Q's and ΔS's have the same numerical values as for the reversible heat engine, but they have opposite signs. If the heat pump is irreversible, then more work must be done on the system for the same Q_C removed from the cold reservoir. As a result, Q_H must be larger than before, and for irreversible operation we must have

$$\frac{Q_H}{T_H} + \frac{Q_C}{T_C} = \Delta S_H + \Delta S_C = \Delta S_{\text{total}} > 0$$

Thus for the heat pump as well as for the heat engine we may write

Reversible	*Irreversible*
$\Delta S_{\text{total}} = 0$	$\Delta S_{\text{total}} > 0$

These results for thermal processes are the same as those obtained for the mechanical processes considered earlier, and we may summarize all results for both types of process by the equation

$$\Delta S_{\text{total}} \geqq 0$$

where the equality applies to the limiting process called *reversible*, and the inequality to all *irreversible* processes. The two types of processes considered were very different in character, but in spite of this, the same equation has been found to apply to both. The next step is a bold one; we generalize our conclusions with respect to these two types of processes and *postulate* that this equation is valid for all processes which occur in the world. We cannot prove it, but on the other hand, we cannot delay our generalization until we have examined all conceivable processes. We need the generalization so that we can use it for prediction of the results of processes not yet carried out. We cannot, of course, be absolutely certain that every prediction will be right, but confidence in the generalization grows with every successful prediction. During the past 100 years countless predictions have been made, and not a single one has been wrong. Thus our confidence in the generalization has grown so that now it is virtually unbounded, and we regard the above equation as a law of nature—the Second Law of Thermodynamics.

6

More on the Second Law

In the last chapter I tried to make our ideas about entropy seem plausible. Now I should like to demonstrate something of its usefulness, which results, of course, from the Second Law. This law is a sweeping generalization to the effect that for *any* process the sum of all entropy changes occurring as a result of the process is greater than zero and approaches zero in the limit as the process becomes reversible. Mathematically, we have

$$\Delta S_{\text{total}} \geq 0$$

This generalization was first made by Clausius in 1865. He simply *guessed* that it was right and left it for time to *prove* him right. It is the most general statement of the Second Law, and moreover it is quantitative.

Now let's go back to see whether our friend's "box of tricks" can work. To do this, we will need to calculate some numbers for entropy changes. This is done most simply for ideal gases, and there is no need to complicate this illustration by going further. For ideal gases we know two things. The equation of state for 1 mole is $PV = RT$, and the internal energy change is given by $dU = C_V\,dT$. We will need both of these equations.

We are now interested in the *properties* of an ideal gas, and not in any particular *process* which changes these properties. Assume for the moment that the properties of 1 mole of an ideal gas are changed differentially in a reversible process. By the first law

$$dU = dQ_{\text{rev}} - dW_{\text{rev}}$$

But $dQ_{\text{rev}} = T\,dS$ and $dW_{\text{rev}} = P\,dV$. For an ideal gas $dU = C_V\,dT$. Thus

$$C_V\,dT = T\,dS - P\,dV$$

or

$$T\,dS = C_V\,dT + P\,dV$$

But

$$PV = RT \qquad \text{and} \qquad P\,dV + V\,dP = R\,dT$$

or

$$P\,dV = R\,dT - \frac{RT}{P}\,dP$$

Thus

$$T\,dS = C_V\,dT + R\,dT - RT\frac{dP}{P} = (C_V + R)\,dT - RT\frac{dP}{P}$$

or

$$T\,dS = C_P\,dT - RT\frac{dP}{P}$$

and

$$dS = C_P \frac{dT}{T} - R \frac{dP}{P}$$

If C_P is constant, we may integrate to get

$$\Delta S = C_P \ln \frac{T_2}{T_1} - R \ln \frac{P_2}{P_1}$$

and we have an equation for the calculation of the entropy changes of an ideal gas with constant heat capacities. This derivation illustrates the theoretical half of thermodynamics, useful for the calculation of property changes regardless of the process causing them. We will now use this result in the applied half of thermodynamics to calculate the entropy changes for the process claimed by our friend with the box of tricks.

To do calculations, we need quantitative information; so we press our friend for specifics as to what his device can do. He says compressed air at 4 atm and 70°F will produce two equal streams at 1 atm, one at 0°F and the other at 140°F. Thus,

1 lb mole air at 1 atm, 0°F ← Box of tricks → 1 lb mole air at 1 atm, 140°F

↑

2 lb moles air at 4 atm, 70°F

We may consider the process in two parts. One mole of air changes from 4 atm and 70°F to 1 atm and 0°F. Taking $C_P = 7$ Btu/lb mole–°F for air, we calculate ΔS for this change to be

$$\Delta S_{\mathrm{I}} = C_P \ln \frac{T_2}{T_1} - R \ln \frac{P_2}{P_1} = 7 \ln \frac{460 + 0}{460 + 70} - 2 \ln \tfrac{1}{4}$$
$$= 1.79 \text{ Btu/lb mole–°R}$$

Similarly, for the other mole of air we have

$$\Delta S_{\mathrm{II}} = C_P \ln \frac{T_2}{T_1} - R \ln \frac{P_2}{P_1} = 7 \ln \frac{460 + 140}{460 + 70} - 2 \ln 1/4$$
$$= 3.65 \text{ Btu/lb mole–°R}$$

$$\Delta S_{\text{total}} = \Delta S_{\mathrm{I}} + \Delta S_{\mathrm{II}} = 5.44 \text{ Btu/°R} > 0$$

This result shows not only that the process can work, but that there's lots of room to spare. After all, ΔS_{total} need only be differentially greater than zero for the process to be possible. Here it is much greater than zero, indicating that the process is highly irreversible and hence easy to make work. Note that we did not know what was *in* the box of tricks to determine whether it *could* work. And that's all we have decided—that it is *possible*. Whether the man's particular device *will* work depends on his ingenuity, that is, on how good an engineer he is. Thermodynamics merely puts a limit on genius.

It is also true that our friend could have claimed too much for his device. In the first place he could have proposed conditions that violate the First Law, but he did not. His claim was that the cold-air stream dropped in temperature by 70°F and that the hot-air stream rose by 70°F. This is entirely consistent with his claim that equal amounts of air exist in both streams and that the apparatus is well insulated. But he could also have claimed conditions that violate the Second Law. For example, he might have said that with 2 moles of compressed air at 70°F and 1.1 atm he could produce 1.6 moles of air at 0°F and 0.4 mole at 350°F, both at atmospheric pressure; thus

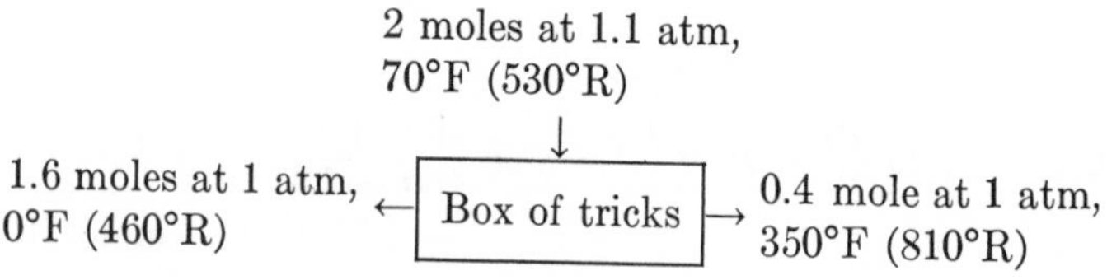

Again, these figures are consistent with the First Law, for the cold stream is 4 times larger than the hot stream, but its temperature change is only one-quarter as much. We calculate entropy changes of the two streams as before; thus

$$\Delta S_{\text{I}} = m_{\text{I}}\left(C_P \ln \frac{T_2}{T_1} - R \ln \frac{P_2}{P_1}\right)$$

$$= 1.6\left(7 \ln {}^{460}\!/_{530} - 2 \ln \frac{1}{1.1}\right) = -1.28 \text{ Btu/°R}$$

$$\Delta S_{\text{II}} = m_{\text{II}}\left(C_P \ln \frac{T_2}{T_1} - R \ln \frac{P_2}{P_1}\right)$$

$$= 0.4\left(7 \ln {}^{810}\!/_{530} - 2 \ln \frac{1}{1.1}\right) = +1.27 \text{ Btu/°R}$$

$$\Delta S_{\text{total}} = \Delta S_{\text{I}} + \Delta S_{\text{II}} = -0.01 \text{ Btu/°R} < 0$$

The Second Law denies that this result can be realized. Thus we see the need to have a quantitative expression for the Second Law. Many types of processes are possible only within limits, and we now have a simple way to determine the limits.

We might consider *how* such a device could work, and it is not difficult to devise a means. If the compressed air is run through a turbine, the turbine produces work at the expense of the internal energy of the air. Thus the temperature of the air drops as it expands to atmospheric pressure in the turbine. This air stream may be split into two parts

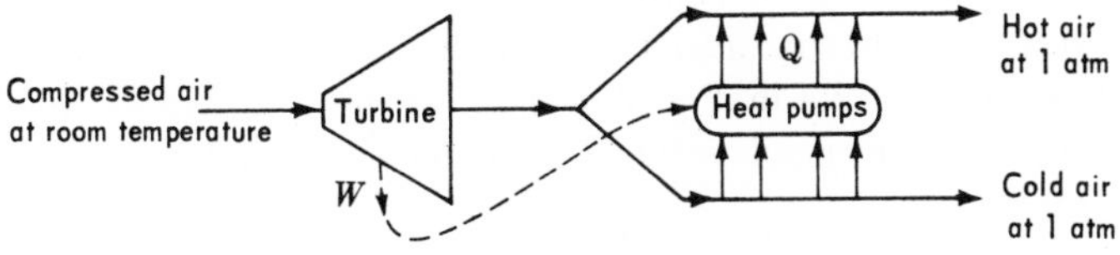

Figure 6-1

as shown in Fig. 6-1. The work output of the turbine can be used to operate refrigerators or heat pumps which extract heat from one stream and discharge heat into the other.

The result is a hot-air stream and a cold-air stream. We can even imagine this process being carried out reversibly. Clearly, the process is possible not only from a thermodynamic but also from a mechanical point of view.

However, our friend claims his device has no moving parts, and that is the only feature which now seems remarkable. As a matter of fact there *is* such a device, and it is known as the *Hilsch-Ranque vortex tube.* It was first invented in the 1930s by a Frenchman named Ranque, but no one paid any attention. It was not until the Allies entered Germany after World War II that these devices became known. During the war a German by the name of Hilsch had built a number of them and distributed them in German laboratories. Each consisted of a straight piece of tubing or pipe which was divided into two parts by an orifice as shown in Fig. 6-2.

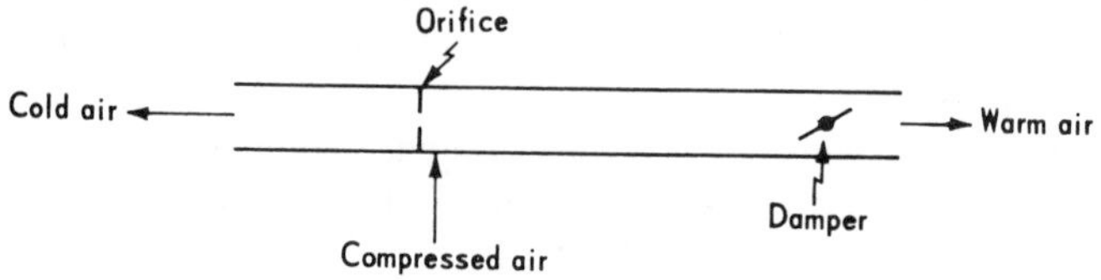

Figure 6-2

Compressed air is fed in through a connection adjacent to the orifice, but it is introduced in a special way, that is, tangentially to the tube. This is shown by the cross-sectional view of the tube in Fig. 6-3. This entering air, expanding into the tube, reaches fairly high velocities and imparts a

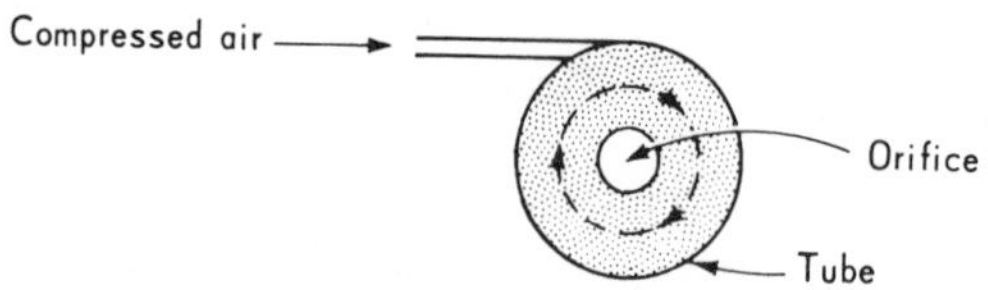

Figure 6-3

rotary motion to the air. This is the reason it's called a *vortex* tube. In expanding into the tube from a high pressure to a low pressure, the air has gained kinetic energy at the expense of its internal energy. Thus the temperature of the rotating air mass is lower than the initial air temperature, and it is more or less uniform over the cross section of the tube. However, the linear velocity, and therefore the kinetic energy of the air, varies from place to place in the air vortex, which rotates more or less as a unit. Thus air at the center of the tube has a very low velocity and kinetic energy, and these increase as we move outward from the center toward the wall. Very near the wall this analysis fails, but this does not destroy the general line of reasoning. The air near the center of the tube passes through the orifice and leaves the tube as the cold air stream. The air in the outer parts of the tube is also at a low temperature, but this air has a high kinetic energy. As it swirls down the tube away from the orifice, it loses its kinetic energy as a result of turbulence, internal friction, and mixing, and this kinetic energy reverts to internal energy. Thus the gas temperature rises, and the gas comes out of the tube hotter than the initial temperature because nearly all of the kinetic energy generated by expansion reappears as internal energy in only *part* of the initial gas stream. The highly irreversible nature of this process is clear, and this is the reason that data taken on actual tubes always show large increases in entropy.

The sole attractive feature of this device from a practical point of view is the absence of any moving parts. Thus the devices are cheap to build and require no maintenance, and they become practical devices only where these factors are of overriding importance. Thus we find them in use to cool drinking water on some railroad locomotives. Union work rules require that cool water be available for the crew, and unless it is, the locomotive is out of service. Downtime on locomotives is so expensive that it is a big advantage to have a foolproof water cooler. All that is needed is compressed air, and no train can operate anyway without that, for if the air fails, the brakes lock.

After this digression we return to the Second Law of Thermodynamics. In our expression of this law,

$$\Delta S_{\text{total}} \geqq 0$$

the word "total" is meant to imply that we take into account all changes in both the system and its surroundings, and we can equally well write

$$\Delta S_{\text{system}} + \Delta S_{\text{surroundings}} \geqq 0$$

My purpose in writing this alternative expression of the Second Law is to make clear that the Second Law applies to the system and its surroundings taken together and does not impose any general restraint on the system alone. This is also true of the First Law, which may be written

$$\Delta E_{\text{system}} + \Delta E_{\text{surroundings}} = 0$$

where E represents energy in general.

There is a special kind of system, called an *isolated* system, which is completely cut off from its surroundings; that is, it can exchange neither energy nor matter with its surroundings. Thus changes which occur in such a system cannot cause changes in the surroundings, and we need consider just the system and no more. In this case our equations become

$$\Delta S_{\text{system}} \geqq 0$$

$$\Delta E_{\text{system}} = 0$$

For such a system the total energy must remain constant, and the total entropy can only increase or stay constant. If the system is in *equilibrium*, its properties, including the entropy, do not change. Thus the equality $\Delta S_{\text{system}} = 0$ implies that equilibrium has been reached, and the inequality $\Delta S_{\text{system}} > 0$ implies a change toward equilibrium. Since the entropy of the system can only *increase*, this must mean that the equilibrium state is that state which produces the *maximum possible* entropy. Thus at equilibrium the entropy of an isolated system has its maximum value with respect to all possible variations, and the condition for this maximum is that $dS_{\text{system}} = 0$. We will now apply this condition to a simple situation.

Imagine a cylinder closed at both ends and containing a piston which divides the volume of the cylinder into two parts as shown in Fig. 6-4. We imagine the cylinder to be a

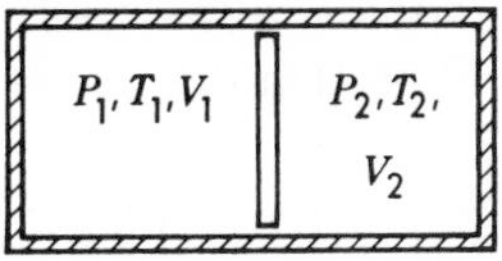

Figure 6-4

perfect insulator and the piston to be a heat conductor. We also imagine the piston to move in the cylinder without friction. The question is how P_1 and P_2 are related and how T_1 and T_2 are related when the system is at equilibrium. The answer is, of course, obvious; $P_1 = P_2$, and $T_1 = T_2$. But our problem is to see whether our criterion of equilibrium $dS_{\text{system}} = 0$ will in fact predict these results.

To do this, we need the fundamental property relation

$$dU = T\,dS - P\,dV$$

We may apply this equation separately to the two parts of our system on the left and right of the piston. We take U, S, and V to mean the total properties of all the gas in whichever part of the system we are considering. Thus the property relation may be solved for dS and written for each part of the system as follows:

$$dS_1 = \frac{dU_1}{T_1} + \frac{P_1}{T_1}\,dV_1$$

$$dS_2 = \frac{dU_2}{T_2} + \frac{P_2}{T_2}\,dV_2$$

Adding, we obtain

$$dS_1 + dS_2 = d(S_1 + S_2) = dS_{\text{system}}$$
$$= \frac{dU_1}{T_1} + \frac{dU_2}{T_2} + \frac{P_1}{T_1}\,dV_1 + \frac{P_2}{T_2}\,dV_2$$

Since the system is isolated, its energy is constant, and

$$dU_1 + dU_2 = 0 \qquad \text{or} \qquad dU_2 = -dU_1$$

Furthermore, the total volume of the system is constant, and we have, in like fashion, $dV_2 = -dV_1$. We also impose our equilibrium condition that $dS_{\text{system}} = 0$, and obtain

$$dS_{\text{system}} = \left(\frac{1}{T_1} - \frac{1}{T_2}\right) dU_1 + \left(\frac{P_1}{T_1} - \frac{P_2}{T_2}\right) dV_1 = 0$$

The question now is how this equation can be identically zero for all imaginable variations dU_1 and dV_1. These variations are independent and arbitrary, for we can easily imagine small heat flows through the piston which cause the change dU_1, while at the same time the piston moves slightly to cause a noncompensating change dV_1. We conclude that each term of the equation must separately be zero, and this can be generally true only if

$$\frac{1}{T_1} - \frac{1}{T_2} = 0 \qquad \text{and} \qquad \frac{P_1}{T_1} - \frac{P_2}{T_2} = 0$$

From these we see immediately that for equilibrium we must have $T_1 = T_2$ and $P_1 = P_2$, results which we know intuitively to be correct.

The point of this trivial example is to show that our criterion of equilibrium involving the entropy can be used to produce meaningful results. It can also be applied to nontrivial problems. For example, what if the cylinder had contained no piston but had in it a reactive mixture of gases? What intuitive notion will tell you the equilibrium composition? The long and short of it is that one important use of thermodynamics is in the prediction of equilibrium states.

The equation $\Delta S_{\text{total}} \geqq 0$ has been presented as the most general statement of the Second Law. Let us now see how it stands in relationship to the other statements so often given of this law:

1. "No engine, operating in a cycle, can convert all of the heat it takes in into work." Such an engine would produce but one entropy change, that caused by removing heat from a heat reservoir. This is an entropy *decrease*. Thus we would have $\Delta S_{\text{total}} < 0$, whereas the Second Law requires $\Delta S_{\text{total}} \geqq 0$.
2. "Heat cannot be caused to flow from a cooler to a hotter body without producing some other effect." If the excluded process were possible, we would have heat extracted from a body at T_C, causing an entropy change $-Q/T_C$, and added to a body at T_H, causing an entropy change Q/T_H. Then we would have

$$\Delta S_{\text{total}} = \frac{-Q}{T_C} + \frac{Q}{T_H} = Q\left(\frac{1}{T_H} - \frac{1}{T_C}\right) = \frac{Q(T_C - T_H)}{T_H T_C}$$

Since $T_C < T_H$, this gives $\Delta S_{\text{total}} < 0$, which is contrary to our statement of the Second Law.

Thus we see that these negative statements appear as consequences of our positive statement with respect to entropy changes and are hardly reasonable alternatives to it.

I would make one further point. In standard classical treatments of the Second Law, the Carnot engine cycle appears to be of overwhelming importance, and students get the idea that the Carnot engine is the beginning and the end of the Second Law. Actually, in making the generalization necessary to reach the Second Law, we could start with most any process. The heat engine was employed by the founding fathers because of its great interest and value. Such engines are still of tremendous technical importance, and I have paid considerable attention to them for that reason. However, in the application of thermodynamics to engineering devices, the heat engine is just one more device, and the equations for it represent just another example of the utility of the laws of thermodynamics. Thus for a heat engine operating reversibly between two heat reservoirs at T_H and T_C, we have from the First Law that

$$Q_H + Q_C - W = 0 \qquad \text{or} \qquad W = Q_H + Q_C$$

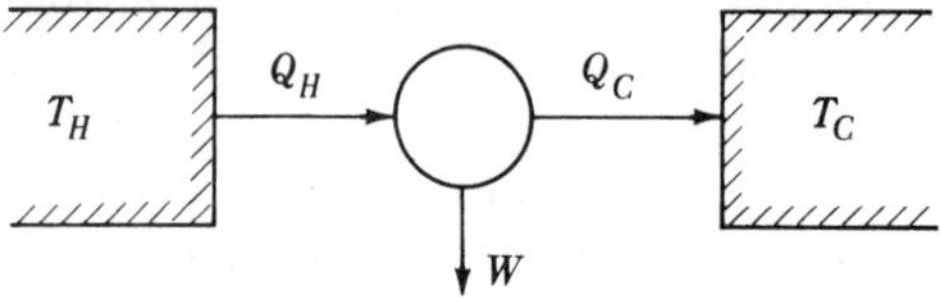

Figure 6-5

and from the Second Law that

$$\Delta S_{\text{total}} = \frac{Q_H}{T_H} + \frac{Q_C}{T_C} = 0$$

Thus

$$Q_C = -Q_H \frac{T_C}{T_H}$$

and

$$W = Q_H + Q_C = Q_H - Q_H \frac{T_C}{T_H} = Q_H \left(1 - \frac{T_C}{T_H}\right)$$

Finally,

$$\eta = \frac{W}{Q_H} = 1 - \frac{T_C}{T_H}$$

Thus there is no special need for you to remember the Carnot engine formula. The two basic laws cover this process and all others besides.

7

Thermodynamics and Statistical Mechanics

It has not been necessary in our discussions of thermodynamics to mention the nature of matter, nor is it *necessary* now. As far as classical thermodynamics is concerned, matter may as well not be made up of atoms. But our belief in atoms and molecules is pretty firm, and we gain nothing by ignoring the atomistic nature of matter. It makes much more sense to ask what can be added to thermodynamics by knowing something of the structure of matter. This type of inquiry has led to the development of kinetic theory and statistical mechanics. It is sometimes said that the development of thermodynamics preceded the development of those subjects which rely on the atomic nature of matter. But this is hardly true, for to a considerable degree they developed simultaneously, often in the minds of the same individuals. The first book on applied thermodynamics was published by Rankine in 1859, the same year that Maxwell published his first paper on the dynamical theory of gases. Thermodynamics, kinetic theory, and statistical mechanics after 1850 grew up together and eventually led to the quantum theory. It is often forgotten that Max Planck took thermodynamics and statistical mechanics as his special fields of interest, and it was difficulties that arose in these fields that led him to the postulate that energy is quantized.

Thus questions about the interrelation between thermodynamics and molecular behavior arose very early, and perhaps the most famous problem of this nature was posed by Maxwell in 1871 under the heading "Limitations of the Second Law." He invented (in his mind) a *being* that could deal directly with molecules; this being has since been known as Maxwell's demon. Maxwell suggested that a container filled with gas be divided by a partition in which there was a trap door manned by his demon. The demon would observe molecules approaching the trap door from both sides and would operate the door so as to allow only fast molecules to pass in one direction and only slow molecules to pass in the other direction. Thus the demon would act to sort molecules according to speed. As a result the gas on one side of the partition would become increasingly warmer and that on the other side cooler. At some point we could start a heat engine to operate between the two temperatures, and it could deliver work continuously to the surroundings as long as the demon continued his activities. It would only be necessary to add heat to the system to compensate for the work done, and we would have an engine operating in a cycle that converted all the heat taken in into work, in violation of the Second Law. Maxwell imagined that his demon itself did no work; it was a reversible demon that released energy to open a frictionless trap door and recovered the same energy when the door closed.

There have been many suggestions as to how to build a device to violate the Second Law, but not one has ever been demonstrated to do so. However, Maxwell added a new dimension to this endeavor. He postulated a being, intelligent in some sense, that could deal with individual molecules. The easiest way out is to declare that the Second Law denies that such a being could exist. But life has always been mysterious, and we inevitably suppose that it must have qualities not fully taken into account by the known laws of physics. After all, bacteria are very small beings whose accomplishments are by no means inconsequential. So Maxwell's demon has not been lightly dismissed, and even after almost 100 years it is still a fascinating topic of discussion. I might remark that a demon that sorts molecules according to speed is not the only demon one can imagine. Sorting on molecular species is another example, but the problem with respect to the Second Law is no different.

Maxwell's demon appears to violate the Second Law through its ability to deal with individual molecules. Is this the key to "success," or are there other ways to take advantage of the molecular nature of matter in efforts to violate the Second Law? Let me pose another problem. Suppose again we have a container divided by a partition. This time there is gas on one side of the partition, but a total vacuum on the other. The partition is removed and the gas expands to fill the total volume. Now the question is whether the gas will ever of its own accord return to its initial location in one part of the container. The overwhelming consensus of informed opinion is that it will, provided one waits long enough! It comes down to a matter of chance. Since the gas molecules are in continual motion, one concludes that there is a finite (but minuscule) probability, a chance, at any instant that the original configuration will be reproduced. One need not even insist on a special initial configuration. It is sufficient to consider the container merely filled with gas and then to ask whether the gas will ever momentarily collect itself in any portion of the container. For if it does, then we can insert a partition and trap the gas in a state of lower entropy than it had initially. Again the overwhelming consensus of informed opinion is that this is possible if one is prepared to wait long enough, say, $10^{10^{10}}$ years. Whether or not you believe this will ever happen is not important. We can imagine it to happen regardless of whether it actually will, and whether real or imaginary, it represents a process seemingly at odds with the Second Law, and one that does not require dealing with individual molecules. It is important to note, however, that to accomplish the process one must insert a partition into the container at exactly the right instant. Thus the process does require continuous observation of the system by some being or device capable of detecting molecules and taking appropriate action. So we see that the two hypothetical processes just described do have common elements and need not be considered independently.

In particular, they have in common the idea of molecular ordering. In the case of Maxwell's demon ordering is done on the basis of speed; that is, high-speed molecules are segregated from low-speed molecules. In the second case, ordering is done in the sense that molecules are collected from a larger region of space into a more restricted region. This process, by the way, could also be accomplished by a Maxwell demon, one which allowed molecules to pass only one way through its trap door. Another type of ordering process that a Maxwell demon could accomplish results when a gas *mixture* is admitted to the container. The demon could operate his trap door so as to allow green molecules to go only one way and red molecules only the other. This would serve to segregate the green from the red molecules. So you see that our use of the word "ordering" gives it perhaps a broader meaning than is found in its everyday use. Our demon is said to bring about ordering whenever he restricts molecules to a given region of space by virtue of some characteristic of the molecule. We have considered molecular speed, molecular species, and even the very characteristic of being a molecule at all. Molecules left to themselves do not become so ordered except, as we have seen, by chance. Ordering at will requires the intervention of some outside agent, of which Maxwell's demon is a very special example.

All of these ordering processes produce a *reduction* in the entropy of the system, and each reduction can easily be calculated by the methods of thermodynamics. We are therefore led to the notion that increasing order corresponds to decreasing entropy, and vice versa; this is the basic idea that underlies statistical mechanics. All that we need in addition is a method of expressing order or disorder in a quantitative way, but this we will leave for later.

There is another aspect of the sorting processes involving Maxwell demons that we have not yet considered. It centers around the fact that the demon must act on the basis of *information;* that is, he cannot act properly until he knows that a molecule in a particular place is a fast one, a slow one, a red one, a green one, directed left, or directed right, etc. Even a demon that sits around waiting for chance to order a system must keep continuously informed of the locations of molecules; otherwise he would never know the moment to insert a partition so as to preserve the long-awaited but otherwise-momentary order. There are two separate ideas which come out of these observations. The first is that there may be some connection between entropy and information. The second is that the information-gathering activities of the demon may be the key to whether or not he can operate so as to cause violations of the Second Law. The apparent link between information and entropy has been exploited and developed into the subject called *information theory*, which has important applications in the design of communication systems. The fundamental equation of information theory is identical with the equation for entropy in statistical mechanics, and the quantity calculated, having to do with the information content of messages, is even called *entropy*. In statistical mechanics we deduce the properties of matter by applying statistics to large numbers of molecules. In information theory we deduce the information-carrying capacity of communications systems by applying statistics to large numbers of messages.

In our descriptions of the activities and ambitions of Maxwell demons we have implied several questions. Let me state these questions one by one, and provide what are thought to be correct answers:

1. Is it necessary to regard the demon as a *living* being? The answer is that it is not *necessary*. Moreover, it's not even advantageous. The demon is merely an intermediary, a relay mechanism, that responds in a specific way to an information signal. It may therefore be automated or programmed to perform its tasks at least as surely as if it possessed the intelligence of a human being. This is not to deny that an intelligent living being could serve as a sorting demon, but such a demon would be at a disadvantage. In spite of the mysteries of life, every study of life processes has demonstrated that the laws of physics *do* in fact apply. There may be *additional* laws, but none of those known is violated, not even the Second Law of Thermodynamics. The fantastic ordering of atoms and molecules necessary to produce and maintain a living system is accompanied by a more than compensating disorder created in the surroundings. Thus the ever-increasing order represented by increasing numbers of the human species is more than matched by the trail of disorder left in our surroundings, of which the increasing pollution of our atmosphere, rivers, and oceans is

but an example. It is not our problem to explain *how* or *why* the ordering necessary to living systems occurs. The fact is that it does, and it does so without violating the laws of physics. So the attribution of life to Maxwell's demon can only prejudice the case against its ambitions to violate the Second Law. We can therefore narrow our attention to automated devices.

2. Can an automated device be activated by mechanical interaction with the molecules themselves? The answer to this is no, for the following reason. Any device or portion of a device used to trigger the necessary action to accomplish the sorting of molecules can be no more massive than the molecules themselves; otherwise it would grossly interfere with the motions of the molecules and actually prevent the sorting from being accomplished. On the other hand, any object no more massive than the molecules will be subject to the same thermal motions as the molecules, and as a result cannot be held in its proper location. We are therefore reduced to more massive devices which rely on *information* about the molecules to be sorted, and this can be transmitted only by electromagnetic radiation, of which ordinary light is one example.
3. Can an automated device relying on information sort molecules in violation of the Second Law? Again, the answer is that it cannot. We may presume our device to be sufficiently massive and to be mechanically reversible, and we concentrate on the problem of how it is to sense the molecules with which it is designed to deal. The device is enclosed within a container, and the only way it can sense its subject molecules without grossly disturbing their motions is by some form of electromagnetic radiation. Thus the device must radiate energy and sense molecules through their reflection of radiation. This energy is absorbed throughout the system and can be shown to cause a greater entropy increase

than any decrease caused by the proper working of the device.

Thus we conclude that Maxwell's demon cannot operate in such a way as to violate the Second Law of Thermodynamics. The fact that this problem has been kicked around for almost 100 years and is still of interest illustrates the reluctance with which even scientists accept the Second Law as being inviolate. There is good reason for *wanting* to violate it, for if it could be done, all of man's energy requirements could forever be met without any depletion of resources or pollution of his surroundings. Man lives on hope, and does not readily take to restrictions on what he can do. So far, however, he has had to live within the limits defined by the laws of thermodynamics, and all indications are that he will continue to.

My purpose in this discussion has been to illustrate the problems encountered when we merely contemplate dealing with individual molecules. Nevertheless, we would like to be able to use our knowledge of the molecular or microscopic nature of matter to help us understand the macroscopic behavior of matter. The enormous numbers of molecules that make up macroscopic systems and the chaotic motions of these molecules suggest that some sort of statistical treatment might prove useful. This idea led to the development of statistical mechanics, a subject so intimately linked with thermodynamics that it seems essential to devote some time to it.

The obvious way to go about developing this subject is to apply statistics to the properties of molecules themselves. However, no such treatment can possibly be general. The reason for this is that molecules interact, and as a result statistical averaging of their *private* properties does not provide any meaningful quantity descriptive of a macroscopic system. For example, molecules do possess their own private kinetic energies, but not their own private potential energies, because potential energy arises through forces acting between molecules and is shared among them. Thus statistical averaging of the private (kinetic) energies of molecules does not in general allow calculation of the internal energy of the system. Only for ideal gases can one say that molecules are independent of one another. I want to avoid at all costs any treatment that is limited to ideal gases. Actually, the cost is negligible, for it is hardly more difficult in statistical mechanics to be general than to be narrow.

If we are not to deal with individual molecules, what is the alternative? It is to deal with a very large collection of macroscopically identical systems known as an *ensemble.* The word ensemble is most commonly applied to a collection of musicians, but the word has meaning only if all the musicians are playing the same tune. The thing that makes a collection of systems an ensemble is that all members would appear to an outside observer to be identical. If we have a closed (i.e., constant-mass) system containing N identical atoms or molecules and having a fixed volume V and existing in thermal equilibrium with a heat reservoir at temperature T, then we regard the thermodynamic state of the system to be fixed, regardless of what the atoms or molecules inside may be doing. The system, far from being restricted to ideal gases, may be solid, liquid, or gas.

We imagine this system to be reproduced or replicated a tremendous number of times, and we imagine this collection of macroscopically identical systems to be arranged on a lattice so that the members are in close contact with one another. Then we imagine the entire collection to be *isolated* from its surroundings; that is, we imagine the boundaries of the collection to be impervious to both the passage of matter and of energy. This is our ensemble. But what is its purpose? It is merely an aid to our mental processes, for we now ask how the various members of the ensemble differ from one another at the *microscopic* or molecular level.

We know that molecules move about in a chaotic fashion and that at any instant the particular molecular configuration to be found in any ensemble member is just a matter of chance. If we consider all the members at any one instant, we therefore expect to find a tremendous variety of microscopic configurations. These configurations, seen at any one instant in the ensemble, are presumed to be the same as those we would see in the original real system were we to observe it for a very long time. Furthermore, we assume that the observed macroscopic properties, such as pressure and internal energy, are averages resulting from the various configurations considered, either over a long period of time or over a very large ensemble.

The question yet to be answered is how the various microscopic configurations are to be characterized. The answer is provided by quantum mechanics, and we must here merely accept it. Quantum theory postulates that energy on the microscopic scale is made up of discrete units or *quanta.* Since energy is quantized, the internal energy of a macroscopic system at any instant is the sum of an enormous number of quanta of energy, and because of this, a macroscopic system at any instant is in a particular *quantum state,* characterized by a particular value of its energy, E_q. There is a discrete set of possible energy values, and we will use the notation $\{E_q\}$ to represent the entire set. Any one value E_q of the set $\{E_q\}$ represents the particular energy associated with quantum state q of the system. The set is discrete because one can never get a complete spectrum of values by summing quanta, just as one cannot obtain decimal numbers by adding integers.

Quantum theory also provides the result that *for a closed system of N particles, the set of values $\{E_q\}$ is completely determined by the volume of the system.* Since we have specified the volume V of our system, we have fixed the set $\{E_q\}$, and we can expect the energy of our system to pass through all these possible values over a long period of time and, in fact, to pass through some of them many, many times. Similarly, our ensemble at a given instant is made up of members each in its own quantum state with an energy E_q taken from the set $\{E_q\}$. There may be many members with the same value of E_q. The *total* internal energy of the entire ensemble is just the sum of the energies E_q, each multiplied by the number of members n_q in the particular quantum state q; thus

$$\mathcal{U} = \sum_q n_q E_q = \text{constant} \tag{7-1}$$

where the summation is over all possible quantum states. The constancy of $\mathcal{U}$ results from the First Law of Thermodynamics as applied to the ensemble, which you will recall is an isolated system. Furthermore, if there are a total of n members in the ensemble, then

$$n = \sum_q n_q = \text{constant} \tag{7-2}$$

The constancy of n again results because the ensemble is considered isolated. These two equations express the restraints on the ensemble, and we will need to take them into account later.

We now want to consider the makeup of our ensemble, or more precisely the number of *different ways* it can be made up. It is here that statistics enters, and the procedure seems strange indeed. It is probably best explained by taking a specific example that simulates the real situation, but on a very small scale. Assume we have an ensemble that contains 24 members; that is, $n = 24$, as shown in Fig. 7-1. Consider for a moment the shaded member of this ensemble. We will identify it with the letter a. It is a replica of our original system, which you will recall was specified to have the volume V and to be in equilibrium with a heat reservoir at temperature T. Member a also has volume V and is in equilibrium with the rest of the ensemble, which then constitutes a heat reservoir at T for member a. Now each of the 24 ensemble members can be identified by its own letter. It is not the positions on the lattice that are being

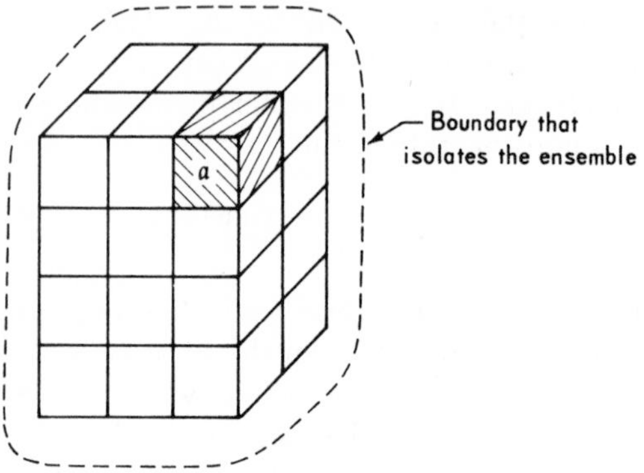

Figure 7-1

identified by letters, but the ensemble members themselves. Thus Fig. 7-1 shows member a in just one of its 24 possible locations, and once it is put in a particular location there are $n - 1 = 23$ positions left for placing member b. Moreover, for *each* of the 24 locations for a there are 23 locations for b. Thus there are 24×23 different ways to locate both a and b. Once a and b are placed, there are 22 ways to locate c for each of the 24×23 locations of a and b. Thus there are $24 \times 23 \times 22$ ways to locate a, b, and c. And so it goes; by the time we have located all $n = 24$ members, we have chosen one arrangement of the ensemble out of $n! = 24! \approx 6.2 \times 10^{23}$ possible arrangements. Clearly, one does not require a large ensemble in order to generate large numbers.

Each member of the ensemble must at any instant be in a particular quantum state characterized by a particular value of E_q taken from the set of possible values $\{E_q\}$. In our example we will assume that there are only four possible quantum states, identified by setting q equal to 1, 2, 3, or 4. Thus $\{E_q\}$ is made up of the energy values E_1, E_2, E_3, and E_4. Each of our 24 ensemble members must be in one of these four quantum states; so let us assign a quantum state to each ensemble member as indicated in Table 7-1.

Table 7-1

Quantum state, q	*Energy, E_q*	*Ensemble members in state q*	*Number of members, n_q*
1	E_1	$a, b, f, i, l, m, r, v, y$	$n_1 = 9$
2	E_2	d, e, k, o, p, w, x	$n_2 = 7$
3	E_3	c, g, j, s, t, u	$n_3 = 6$
4	E_4	h, z	$n_4 = 2$
			$n = 24$

With this assignment of ensemble members to quantum states, each arrangement of members on the lattice of Fig. 7-1 represents a particular distribution of energy states, which corresponds to a particular sequence in time for the original system to pass through the same states. The complicating factor is that there is more than one ensemble member in each quantum state. If we were to interchange members of the same quantum state, the lattice would look no different. For example, members h and z are both in quantum state 4 and are therefore identical. Interchanging them makes no difference, but our $n! = 24!$ ways of arranging the lattice counted them as different. Thus for *any* arrangement of the lattice we have a second arrangement that is no different, for we may interchange h and z whatever their locations. Thus to get the number of really different lattice arrangements, we would divide $n!$ by $2 = 2! = n_4!$ because of the fact that h and z are identical. Similarly, the six members c, g, j, s, t, and u are all in quantum state 3 and are therefore indistinguishable from one another. There are $6! = n_3!$ ways to arrange these six members on their respective lattice sites without making the lattice any different. This is because c could be put on any of six sites, then g on any of the five remaining sites, j on four, etc. Thus our $n!$ ways to arrange the lattice is too large by a factor of $6! = n_3!$ and must be divided by this factor to remove the indistinguishable arrangements resulting from the fact that six members are all in quantum state 3. By now it should be clear that we must divide $n!$ by $n_q!$ for *each* of the q quantum states. Thus the number of ensemble or lattice arrangements that are really different is given by

$$\omega = \frac{n!}{n_1!n_2!n_3!n_4!} = \frac{24!}{9!7!6!2!} \approx 2.356 \times 10^{11}$$

which is still a big number, though not so large as 24!. It is worth noting at this point that the natural logarithm of ω is not nearly so imposing a number. In fact, $\ln \omega$ is a mere 26.19.

As a result of this example we can now write down a general formula for the number of really different ways an ensemble of n members can be arranged for q quantum states when n_q is the number of members in a particular quantum state; thus

$$\omega = \frac{n!}{\prod_q n_q!} \tag{7-3}$$

where the sign $\prod_q$ signifies the running product of all the factorials $n_q!$.

Clearly, in order to evaluate ω we must know not only n but also the values of all the n_q; that is, we must know the distribution of the ensemble members among the possible quantum states. In our example, used for the purpose of illustration, we assigned values to the n_q's arbitrarily, and had we assigned different values, we would have calculated a different value for ω. For example, we might have set $n_1 = n_2 = n_3 = 0$ and $n_4 = 24$. Then ω would have been 1, and $\ln \omega$ would have been zero. Or we might have set $n_1 = n_2 = n_3 = n_4 = 6$, and ω would have come out to about 2.3×10^{12} and $\ln \omega$ to about 28.5. Evidently, the problem is to find a preferred distribution. The clue to a solution to this problem is found in the Second Law of Thermodynamics.

The ensemble of members that we have devised is isolated from its surroundings. Thus at equilibrium its total entropy must be a maximum with respect to all possible internal variations within the ensemble. Thus we conclude that the preferred distribution of ensemble members among the possible quantum states is that distribution which maximizes the entropy of the entire ensemble. The problem now is to find a connection between entropy and some variable which pertains to the ensemble, and here the best we can do is make an educated guess. We have already suggested that there is some connection between entropy and disorder. Moreover, the quantity ω that we have so painfully developed a formula for in Eq. (7-3) is a measure of the disorder in our original system, for each really different arrangement of our ensemble corresponds to a different sequence of states over a period of time in the original system. The more such possibilities there are, the more chaotic or unpredictable or disordered the original system appears. The limiting case where all ensemble members are in the same quantum state led to a value of $\omega = 1$ or $\ln \omega = 0$. This corresponds to perfect order, for it means that the original system is always in the same quantum state; it never changes its state, and in this sense is in no way chaotic but is completely predictable or ordered. So it makes sense to guess that the total entropy of the ensemble is some function of ω, and all that remains is to find that functional relationship which leads to results that agree with experiment. The relationship which leads to success identifies the ensemble entropy with $\ln \omega$ according to the equation

$$\mathcal{S} = k \ln \omega \tag{7-4}$$

where $\mathcal{S}$ is the entropy of the entire ensemble and k is a constant, known as *Boltzmann's constant.*

Equation (7-4) is the fundamental postulate of statistical mechanics. There is no way to prove it. We can only guess that it is right and then test the consequences against experiment. The remaining steps are mathematical. First, we use Eq. (7-3) to get an expression for $\ln \omega$, and then we simplify this expression as much as possible and substitute the result into Eq. (7-4) to get an expression for the entropy. Finally, we maximize the entropy to obtain the preferred distribution of quantum states. It is done as follows.

Taking the natural logarithm of both sides of Eq. (7-3), we get

$$\ln \omega = \ln n! - \sum_q \ln n_q!$$

Since n is taken to be arbitrarily large for any ensemble of interest, the n_q are also presumed to be large numbers, and in this case we may use Stirling's formula for the logarithms of factorials; thus

$$\ln X! = X \ln X - X$$

Our equation for $\ln \omega$ now becomes

$$\ln \omega = n \ln n - n - \sum_q (n_q \ln n_q) + \sum_q n_q$$

But $n = \sum_q n_q$; therefore

$$\ln \omega = n \ln n - \sum_q (n_q \ln n_q)$$

By a little manipulation we can put this equation into an

even simpler form. First, we factor n; thus

$$\ln \omega = n\left[\ln n - \frac{1}{n}\sum_q (n_q \ln n_q)\right]$$

Multiplying the first term in brackets by $\sum_q n_q/n = 1$, we obtain

$$\ln \omega = n\left[\frac{\sum_q n_q}{n} \ln n - \frac{1}{n}\sum_q (n_q \ln n_q)\right]$$

Now n and $\ln n$ are the same for all terms in the summations, and we may therefore write

$$\ln \omega = n\left[\sum_q \left(\frac{n_q}{n} \ln n\right) - \sum_q \left(\frac{n_q}{n} \ln n_q\right)\right]$$

or

$$\ln \omega = -n \sum_q \left[\frac{n_q}{n} (\ln n_q - \ln n)\right]$$

or

$$\ln \omega = -n \sum_q \left(\frac{n_q}{n} \ln \frac{n_q}{n}\right)$$

We now define the *probability* of quantum state q by

$$P_q = \frac{n_q}{n}$$

So that we have finally

$$\ln \omega = -n \sum_q (P_q \ln P_q) \tag{7-5}$$

Substitution of Eq. (7-5) into Eq. (7-4) gives the following expression for the entropy of the ensemble:

$$\mathcal{S} = -kn \sum_q (P_q \ln P_q) \tag{7-6}$$

We have come now to the problem of finding the set of probabilities $\{P_q\}$ which maximizes $\mathcal{S}$. Unfortunately, this is not the simple maximum problem that it may seem. The reason is that there are two restraints on the system imposed by Eqs. (7-1) and (7-2). However, the problem of finding a maximum subject to restraints has a standard solution through Lagrange's method of undetermined multipliers. Equation (7-1) may be written

$$\sum_q (n_q E_q) = n \sum_q \left(\frac{n_q}{n} E_q\right) = n \sum_q (P_q E_q) = \mathcal{U}$$

or

$$n \sum_q (P_q E_q) - \mathcal{U} = 0$$

We now multiply this equation by an undetermined constant, say λ; thus

$$\lambda \left[n \sum_q (P_q E_q) - \mathcal{U}\right] = 0$$

Since the left side of this equation is zero, it may be added to the right side of Eq. (7-6) without changing anything to get

$$\mathcal{S} = -kn \sum_q (P_q \ln P_q) + \lambda \left[n \sum_q (P_q E_q) - \mathcal{U}\right]$$

This equation now incorporates one of the restraints on the system. To maximize $\mathcal{S}$ subject to this restraint, we differentiate and set $d\mathcal{S} = 0$ (note that k, n, λ, $\mathcal{U}$, and the E_q's are all constant) as follows:

$$d\mathcal{S} = -kn \left[\sum_q (P_q \, d \ln P_q) + \sum_q (\ln P_q \, dP_q)\right] + \lambda n \sum_q (E_q \, dP_q) = 0$$

Thus since $d \ln P_q = dP_q/P_q$, we have

$$\left[\sum_q dP_q + \sum_q (\ln P_q \, dP_q)\right] - \frac{\lambda}{k} \sum_q (E_q \, dP_q) = 0$$

For simplicity, set $\lambda/k = -\beta$ and collect like terms; thus

$$\sum_q (1 + \ln P_q + \beta E_q)\, dP_q = 0 \tag{7-7}$$

Now we impose the second restraint on our system. Dividing Eq. (7-2) by n, we get

$$1 = \sum_q \frac{n_q}{n} = \sum_q P_q$$

Differentiating, we have

$$\sum_q dP_q = 0 \tag{7-8}$$

In order to satisfy both Eqs. (7-7) and (7-8), we must have

$$1 + \ln P_q + \beta E_q = \text{constant}$$

or

$$\ln P_q + \beta E_q = \text{constant} - 1 = A$$

or

$$\ln P_q = A - \beta E_q$$

In exponential form this equation becomes

$$P_q = e^A e^{-\beta E_q}$$

If we now sum the P_q's over all q, we get

$$\sum_q P_q = e^A \sum_q e^{-\beta E_q} = 1$$

Thus

$$e^A = \frac{1}{\sum_q e^{-\beta E_q}}$$

and

$$P_q = \frac{e^{-\beta E_q}}{\sum_q e^{-\beta E_q}} = \frac{e^{-\beta E_q}}{Z} \tag{7-9}$$

where $Z = \sum_q e^{-\beta E_q}$ is called the *partition function.*

The distribution of probabilities for the possible quantum states of our ensemble, as given by Eq. (7-9), is known as the *Boltzmann distribution.* It is by no means the only distribution of probabilities that will lead to the required ensemble energy $\sum_q (n_q E_q)$; it is the particular distribution that maximizes the ensemble entropy and which therefore conforms to the laws of thermodynamics. It is clear from Eq. (7-9) that the only variable on which P_q depends is E_q, because β is constant by definition and Z is a summation that is the same for all the P_q's. Thus the probability of a quantum state for a system of fixed volume in equilibrium with a heat reservoir depends only on the energy E_q of the quantum state, and all quantum states with the same energy have the same probability. We could have used this as an alternative basic postulate of statistical mechanics, and we would have reached the same distribution of probabilities. This course is in fact followed in some textbooks.[1]

Having found the distribution of probabilities, the only remaining question is what to do with it. Since its main use is in the calculation of thermodynamic properties, we should look for equations which give these properties in terms of the variables of statistical mechanics. We start with the two basic expressions, Eq. (7-1) for internal energy and Eq. (7-6) for entropy. If we divide Eq. (7-1) by n, we get the average internal energy of an ensemble member or the time-averaged or macroscopic internal energy of the original system U; thus

$$U = \frac{\mathcal{U}}{n} = \sum_q \left(\frac{n_q}{n} E_q\right)$$

or

$$U = \sum_q (P_q E_q) \tag{7-10}$$

[1] See, for example, M. W. Zemansky and H. C. Van Ness, "Basic Engineering Thermodynamics," Chap. 15, McGraw-Hill Book Co., New York, 1966.

Similarly, if we divide Eq. (7-6) by n, we get the average entropy of an ensemble member or the entropy of the original system S; thus

$$S = -k \sum_q (P_q \ln P_q) \tag{7-11}$$

By Eq. (7-9) we eliminate the P_q's from Eq. (7-10); this gives

$$U = \sum_q \frac{e^{-\beta E_q} E_q}{Z}$$

We also have

$$\left(\frac{\partial \ln Z}{\partial \beta}\right)_V = \frac{1}{Z}\left(\frac{\partial Z}{\partial \beta}\right)_V = \frac{1}{Z}\left(\frac{\partial \Sigma e^{-\beta E_q}}{\partial \beta}\right)_V = \sum_q \frac{e^{-\beta E_q}(-E_q)}{Z}$$

Comparison of the last two equations shows that

$$\left(\frac{\partial \ln Z}{\partial \beta}\right)_V = -U \tag{7-12}$$

Since $Z = \sum_q e^{-\beta E_q}$, we see that Z is in general a function of β and the E_q's. But the E_q's are functions of volume. Thus $Z = Z(\beta, V)$. Therefore

$$d \ln Z = \left(\frac{\partial \ln Z}{\partial \beta}\right)_V d\beta + \left(\frac{\partial \ln Z}{\partial V}\right)_\beta dV$$

or

$$d \ln Z = -U\, d\beta + \left(\frac{\partial \ln Z}{\partial V}\right)_\beta dV$$

But

$$d(U\beta) = U\, d\beta + \beta\, dU$$

or

$$-U\, d\beta = \beta\, dU - d(U\beta)$$

Thus

$$d \ln Z = \beta\, dU - d(U\beta) + \left(\frac{\partial \ln Z}{\partial V}\right)_\beta dV$$

and

$$\beta\, dU = d(\ln Z + U\beta) - \left(\frac{\partial \ln Z}{\partial V}\right)_\beta dV$$

We also have the thermodynamic equation

$$dU = T\, dS - P\, dV$$

Thus

$$\beta T\, dS = \beta\, dU + \beta P\, dV$$

and

$$\beta T\, dS = d(\ln Z + U\beta) + \left[\beta P - \left(\frac{\partial \ln Z}{\partial V}\right)_\beta\right] dV$$

or finally

$$dS = \frac{1}{\beta T} d(\ln Z + U\beta) + \frac{1}{\beta T}\left[\beta P - \left(\frac{\partial \ln Z}{\partial V}\right)_\beta\right] dV \qquad (7\text{-}13)$$

The next step is to develop another general equation for dS. We start by substituting $e^{-\beta E_q}/Z$ for P_q in the logarithm of Eq. (7-11) as follows:

$$\begin{aligned} S &= -k \sum_q P_q \ln \frac{e^{-\beta E_q}}{Z} \\ &= k \sum_q (P_q \ln Z) + k \sum_q (P_q \beta E_q) \\ &= k \ln Z \sum_q P_q + k\beta \sum_q (P_q E_q) \end{aligned}$$

However, $\sum_q P_q = 1$, and by Eq. (7-10), $\sum_q (P_q E_q) = U$. Thus

$$S = k \ln Z + k\beta U = k(\ln Z + \beta U) \qquad (7\text{-}14)$$

and

$$dS = kd(\ln Z + \beta U) \qquad (7\text{-}15)$$

Comparison of Eqs. (7-13) and (7-15) shows that

$$k = \frac{1}{\beta T} \qquad \text{or} \qquad \beta = \frac{1}{kT} \qquad (7\text{-}16)$$

and that

$$\beta P - \left(\frac{\partial \ln Z}{\partial V}\right)_\beta = 0 \qquad \text{or} \qquad P = \frac{1}{\beta}\left(\frac{\partial \ln Z}{\partial V}\right)_\beta \tag{7-17}$$

From Eq. (7-12)

$$U = -\frac{(\partial \ln Z/\partial T)_V}{d\beta/dT} = -\frac{(\partial \ln Z/\partial T)_V}{-1/kT^2}$$

Thus

$$U = kT^2\left(\frac{\partial \ln Z}{\partial T}\right)_V \tag{7-18}$$

and by combining Eq. (7-16) with Eqs. (7-14) and (7-17), we get

$$S = k \ln Z + \frac{U}{T} \tag{7-19}$$

$$P = kT\left(\frac{\partial \ln Z}{\partial V}\right)_T \tag{7-20}$$

Eqs. (7-18) to (7-20) show that the internal energy, the entropy, and the pressure may be calculated once the partition function for a system is known as a function of T and V. Knowing U, S, P, T, and V, we may readily calculate any other thermodynamic property from its definition. Thus statistical mechanics provides a formalism for the calculation of thermodynamic properties from the partition function. Unfortunately, it does not provide the means for the determination of partition functions. This is a problem in quantum mechanics, one that has been solved only for special cases. The relative simplicity of ideal gases allows them to be treated rather completely, and statistical mechanics is widely used for the calculation of the thermodynamic properties of ideal gases from spectroscopic data. Its value here is not that it makes unnecessary the taking of data but that it allows use of a different sort of data (data that are more readily taken) than would be required by classical thermodynamics. For nonideal gases less progress has been made, but statistical mechanics does show that the correct form for an equation of state is the virial form. However, for almost all cases the coefficients in this equation must still be determined from measurements of macroscopic properties.

For liquids, relatively little progress has been made, because one encounters great difficulty in the evaluation of the partition function. For this reason most work on liquids has been directed toward development of approximate methods, none of which is yet regarded as generally satisfactory. Crystalline solids, however, because of their highly ordered state, have been dealt with more successfully. Statistical mechanics has also been applied to the electron gas to provide useful results with respect to the electrical properties of solids. Yet another application is to the photon gas, and this yields important results with respect to radiation. The properties of plasmas, because of their high temperatures, could hardly be determined except by statistical mechanics. All the results of the kinetic theory of gases, such as the Maxwell-Boltzmann distribution of molecular velocities, come out of statistical mechanics; thus kinetic theory as a separate subject is no longer of more than historical interest. I enumerate these applications merely to suggest that you may find a separate study of statistical mechanics to be useful. My purpose here has merely been to show its connection with thermodynamics.

Statistical mechanics adds to thermodynamics on its theoretical side, as a means for or as an aid to the calculation of properties. The other half of thermodynamics, the applied half, benefits only from a wider availability of the data needed in the solution of engineering problems. Although statistical mechanics is based on the presumed reality of atoms and molecules, it does *not* provide, any more than does thermodynamics, a detailed description of atomic and molecular behavior and of atomic and molecular interactions. However, it does provide, as thermodynamics does not, the means by which thermodynamic properties may be calculated whenever detailed descriptions of atomic and molecular behavior are provided from other studies, either theoretical or experimental. Thus statistical mechanics adds something very useful to thermodynamics, but it neither explains thermodynamics nor replaces it.

A CATALOG OF SELECTED

DOVER BOOKS

IN SCIENCE AND MATHEMATICS

A CATALOG OF SELECTED

DOVER BOOKS

IN SCIENCE AND MATHEMATICS

QUALITATIVE THEORY OF DIFFERENTIAL EQUATIONS, V.V. Nemytskii and V.V. Stepanov. Classic graduate-level text by two prominent Soviet mathematicians covers classical differential equations as well as topological dynamics and ergodic theory. Bibliographies. 523pp. 5⅜ × 8½. 65954-2 Pa. $10.95

MATRICES AND LINEAR ALGEBRA, Hans Schneider and George Phillip Barker. Basic textbook covers theory of matrices and its applications to systems of linear equations and related topics such as determinants, eigenvalues and differential equations. Numerous exercises. 432pp. 5⅜ × 8½. 66014-1 Pa. $9.95

QUANTUM THEORY, David Bohm. This advanced undergraduate-level text presents the quantum theory in terms of qualitative and imaginative concepts, followed by specific applications worked out in mathematical detail. Preface. Index. 655pp. 5⅜ × 8½. 65969-0 Pa. $13.95

ATOMIC PHYSICS (8th edition), Max Born. Nobel laureate's lucid treatment of kinetic theory of gases, elementary particles, nuclear atom, wave-corpuscles, atomic structure and spectral lines, much more. Over 40 appendices, bibliography. 495pp. 5⅜ × 8½. 65984-4 Pa. $12.95

ELECTRONIC STRUCTURE AND THE PROPERTIES OF SOLIDS: The Physics of the Chemical Bond, Walter A. Harrison. Innovative text offers basic understanding of the electronic structure of covalent and ionic solids, simple metals, transition metals and their compounds. Problems. 1980 edition. 582pp. 6⅛ × 9¼. 66021-4 Pa. $15.95

BOUNDARY VALUE PROBLEMS OF HEAT CONDUCTION, M. Necati Özisik. Systematic, comprehensive treatment of modern mathematical methods of solving problems in heat conduction and diffusion. Numerous examples and problems. Selected references. Appendices. 505pp. 5⅜ × 8½. 65990-9 Pa. $11.95

A SHORT HISTORY OF CHEMISTRY (3rd edition), J.R. Partington. Classic exposition explores origins of chemistry, alchemy, early medical chemistry, nature of atmosphere, theory of valency, laws and structure of atomic theory, much more. 428pp. 5⅜ × 8½. (Available in U.S. only) 65977-1 Pa. $10.95

A HISTORY OF ASTRONOMY, A. Pannekoek. Well-balanced, carefully reasoned study covers such topics as Ptolemaic theory, work of Copernicus, Kepler, Newton, Eddington's work on stars, much more. Illustrated. References. 521pp. 5⅜ × 8½. 65994-1 Pa. $12.95

PRINCIPLES OF METEOROLOGICAL ANALYSIS, Walter J. Saucier. Highly respected, abundantly illustrated classic reviews atmospheric variables, hydrostatics, static stability, various analyses (scalar, cross-section, isobaric, isentropic, more). For intermediate meteorology students. 454pp. 6⅛ × 9¼. 65979-8 Pa. $14.95